D0894437

The Amoeba in the Room

The Amoeba in the Room

Lives
of the
Microbes

NICHOLAS P. MONEY

OXFORD
UNIVERSITY PRESS

OXFORD
UNIVERSITY PRESS

Oxford University Press is a department of the
University of Oxford. It furthers the University's objective
of excellence in research, scholarship, and education
by publishing worldwide.

Oxford   New York
Auckland   Cape Town   Dar es Salaam   Hong Kong   Karachi
Kuala Lumpur   Madrid   Melbourne   Mexico City   Nairobi
New Delhi   Shanghai   Taipei   Toronto

With offices in
Argentina   Austria   Brazil   Chile   Czech Republic   France   Greece
Guatemala   Hungary   Italy   Japan   Poland   Portugal   Singapore
South Korea   Switzerland   Thailand   Turkey   Ukraine   Vietnam

Oxford is a registered trade mark of Oxford University Press
in the UK and certain other countries.

Published in the United States of America by
Oxford University Press
198 Madison Avenue, New York, NY 10016

© Nicholas P. Money 2014

Library of Congress Cataloging-in-Publication Data
Money, Nicholas P., author.
The amoeba in the room : lives of the microbes / Nicholas P. Money.
pages   cm
Includes bibliographical references and index.
ISBN 978–0–19–994131–5 (hardback)
1. Microbial ecology.   2. Microbiology.   I. Title.
QR100.M647   2014
579'.17—dc23   2013048522

1 3 5 7 9 8 6 4 2
Printed in the United States of America
on acid-free paper

To the memory of Kate Piggott,
my dear grandmother and first biology teacher

And he shewed me a pure river of water of life, clear as crystal
—King James Bible, Revelation 22.1

The revelations of the microscope are perhaps not exceeded in importance by those of the telescope. While exciting our curiosity, our wonder and admiration, they have proved of infinite service in advancing our knowledge of things around us. The present work, founded on such revelations, I have attempted to prepare in a manner to render it easy of comprehension, with the view of promoting and encouraging a taste for microscopic investigation.

—Joseph Leidy, *U.S. Geological Survey of the Territories Report* (1879)

Contents

Acknowledgments

The Amoeba in the Room has been brewing for some time. Kirk Jensen, editor of my first book, was kind enough to support the idea of a project on backyard biodiversity that went well beyond my original fascination with fungi. We spoke about this in New York City, a few days after the tragedy of September 11, 2001, when there was little motivation to contemplate the lives of microbes in ponds. Years passed and thoughts about the unknown diversity of pond life returned with renewed vigor. My memory of Kirk's encouragement helped push this book forward.

I thank my current editors, Latha Menon and Tim Bent, for their work with the manuscript. Archivist Anna Heran provided invaluable help in locating and scanning images from the superb collection of books and journals at the Lloyd Library in Cincinnati. My research collaborator, Mark Fischer, and graduate student, Maribeth Hassett, prepared some of the figures. Artist Debbie Mason (www.debbiemason.com) provided the beautiful drawing of the coccolithophorid alga for Chapter 3. And my wife, Diana Davis, read the chapters and offered editorial suggestions with characteristic insightfulness, patience, and grace.

Preface

I start with an audacious claim: animals and plants are the least part of life. An analogy is helpful to illuminate this assertion, and my late landlord's hairpiece is its subject. "Landy"—as I named him, with the same absence of mind that a child calls his bear "Beary," or crocodile "Snappy"—was a tall hominid of advanced years who favored leopard-print upholstery and snow-white rugs in his immaculately polished home, sited far from the rental properties. He was completely bald, but on his scalp sat a fine-spun confection of blonde wisps. I never figured out how this wig adhered, but it gyrated when he moved his head.

Landy's wig was exactly like earth's large organisms: it was the least important part of him, yet it was the thing that everyone noticed. To attempt comprehension of Landy merely from his wig would be absurd. One would know nothing of his war-hero past, his renown as a naked paraglider, and so on, by examining his well-ventilated nest. And a similarly misplaced purview has been the great failing of biology. All animals, us included, and all plants are an evolutionary afterthought, latecomers to a cosmic game in which the rules were worked out billions of years ago. By looking at the obvious, we have missed, or mostly missed, the essence of life: the amoeba in the room.

In his introduction to A Child's Garden of Verses, published in 1885, Robert Louis Stevenson suggested, "The world is so full of a number of things, I'm sure we should all be as happy as kings." He titled this ditty, *Happy Thought,* and it is one, especially for biologists. There are a lot of different species, nine million or so eukaryotes—organisms whose chromosomes are housed in a nucleus—according to a recent estimate. Animals and plants are eukaryotes, but fewer than one million animals have been catalogued in the last 250 years, and only 200,000 plants are described. Most of the gap between the presumed number of species and the actual number is occupied by microbes.

The prevailing scheme of biological classification refers to four groups, or kingdoms, of miniature organisms: the fungi and the protists, which are eukaryotes, and two kinds of prokaryote—whose cells lack nuclei—called bacteria and archaea. The existence of most of these species was unknown until the brilliant application of the microscope by Robert Hooke and Antonie van Leeuwenhoek in the seventeenth century. The first amoeba wasn't glimpsed until the next century.[1]

There is considerable futility in attempting to count and name all of the species of little things, the "animalcules" of Hooke's era. Here's why. Were a human to attempt to breed with a chimp it is very unlikely that anything would come of it. The same experiment performed with a Shetland pony and the possibility for fertilization recedes further. There is, we accept, unambiguous meaning to the term "species" when we're referring to the cross-sterility of animals. Much the same can be said about plants.

Once we turn to the microscopic organisms, however, the definition of a species becomes more a philosophical question than a scientific one. For bacteria and archaea, especially, the species view of life is almost meaningless. Nevertheless, biologists have named 10,000 or so kinds of these tiny forms of life. Beyond the fact that they don't play by the rules of taxonomy that are used to shoehorn animals and plants into groups, the slimness of this catalogue of germs is due to the biologists' longstanding obsession with big things and related tendency to study the things that are studied most easily. Birds and insects are straightforward: shoot them, or net and poison them, remove their guts or just stick a pin through them, and put them in a drawer. When time permits, open the drawer, and describe the shape and size and color of the animal, and record a host of other features, including, perhaps, some of its activities before it was shot ("It sits on high branches and shits on the heads of naturalists"); write a description and give the creature a Latin name. This has been the practice of taxonomy which has allowed ornithologists to name 10,000 birds and beeologists to identify 20,000 bees.

Other approaches are needed to quantify the reach of microscopic life. An examination of genes, or, more specifically, variations among genes, gets at the foundation of diversity. We can extract DNA from organisms that we have collected, whether or not we have given them a name. And we can isolate their nucleic acids from a sample of seawater, or soil, and figure out what was in the sample without looking at anything with a microscope.

Variations in cell structure can also be useful for examining microbial diversity, but the operation of evolutionary convergence can lead to huge misreadings of kinship where none exist. The filamentous cells of fungi and a group of protists called "water molds" look very similar under the microscope and appear to grow in the same fashion, but these organisms are separated by hundreds of millions of years of evolutionary history. The molecular machinery that separates chromosomes during cell division is a good example of one of the dynamic structures that properly reflects common origins. Close relatives use the same equipment. Analysis of variations in these non-convergent details of cell architecture can be used to make educated guesses about evolutionary histories, and these guesses can be tested by genetics.

The metabolic operation of organisms is a third metric of variation. Plants make their own food and animals eat materials produced by other things. Plenty of microbes operate like animals and plants. All of the fungi and many bacteria consume the tissues grown by other organisms. They are predators and decomposers. Photosynthetic bacteria and algae are primary producers. Marine cyanobacteria, for example, use the sun to cook sugar from atmospheric carbon dioxide just like plants. This common physiology is not an accident of history. Plants do photosynthesis like cyanobacteria because the chloroplasts in plants *are* cyanobacteria (albeit transformed by an ancient endosymbiotic union—one cell swallowing, but not digesting, another cell). Tree genes grow skyscrapers that display solar panels filled with these blue-green bacteria. Microbes have been nourishing themselves in lots of other ways for billions of years. Bacteria and archaea called chemolithotrophs harvest energy from sulfur, ferric iron, and highly reduced inorganic compounds including hydrogen, hydrogen sulfide, ammonia, nitrite, and methane.

Bacteria that energize themselves by oxidizing hydrogen gas have been found in hot springs, and a relative, called *Helicobacter pylori*, resides in the human stomach and may be involved in the development of peptic ulcers and cancer. (On the plus side, the same bacterium seems to aid weight control, and some studies warn that its abolition is linked to the increasing incidence of childhood asthma.) The hydrogen gas that supports these microbes gets spewed into the waters of hot springs by geothermal processes, and into our stomachs by bacteria that live downstream in the intestine. The hydrogen-oxidizing bacteria are among the oldest organisms on earth, and their way of life wouldn't support any of the beings that fill most of the pages of biology textbooks.

Viruses are a crucial part of the picture too, and they are often ignored when biologists wrestle with biological variation. There are reasons for this, none very compelling, including the questionable status of the virus as a non-living entity. In introductory classes in biology, students are taught about the "characteristics" of living things, such as the capacity to reproduce, grow, respond to stimuli, and appreciate fine wines. Viruses reproduce, of course, but they do so inside the cells of host organisms by hijacking the existing biochemical equipment and using it to copy themselves. The formation of each viral particle is not regarded as a growth process, because viruses don't expand like cells—by burning calories and making more of their own substance. And, unlike cells, viruses are incapable of responding to noxious chemicals by swimming away or mounting some kind of physiological defense. In other words, they don't count as organisms when we restrict the appellation to things, like us, that are built from cells: *cellular organisms*.

On the other hand, viruses are assembled from the same kinds of complex biological molecules that constitute cells, and their information is encoded in the same types of nucleic acid. For these reasons, the term *molecular organism* has gained some currency as a catchall for viruses. However they are viewed by biologists, viruses are major players, if not the dominant ones, in life on earth. There are many more viruses than cells. Most of the genetic diversity on our planet comes in the form of viruses.

Viruses, as well as micro-cellular-organisms, present an immense problem for biologists. The smallest forms of life are also an issue for non-biologists, because a failure to comprehend the true nature of biodiversity—the fact that we are soaked in it, filled with it, made of it, and that most of it is invisible—means we lose our sense of place. Because we are steeped in ignorance about the way that most of life works—and would work, and ultimately will work, very well without us—we fail to appreciate the truth, which is that we have been misled by our brains to exaggerate the importance of elephants. We need to employ a little imagination to appreciate the amoeba, and this readjustment holds the only sensible answer to the question about the true meaning of life.

Biology reimagined offers the necessary illumination. So let's begin reimagining. I'll start by considering the great sweep of biological diversity on planet earth, and accomplish this feat without a trip to Yellowstone or an African safari. Indeed, we venture no farther than the small pond and surrounding trees that vibrate with life in my suburban backyard in Ohio. Once understood properly, it is apparent that this plastic-lined oasis matches the biological richness of any National Park. This view of life contrasts with the picture before the seventeenth century, and the revolution in our comprehension of the biosphere made possible by the invention of the microscope. These historical considerations are much more interesting than they sound and occupy the second chapter. Subsequent chapters consider biological diversity in marine ecosystems, in the soil, and in the air. The symbiosis between ten trillion animal cells and 100 trillion microbes that constitutes the human ecosystem is the next topic. Remaining distinctions between *them* (the microbes) versus *us* (cells derived from fertilized eggs) are muddied further by considering the composite nature of the human cell and the viral origin of many of our genes.

Descartes theorized that thinking was proof of existence, but we step beyond biological reality if we imagine that we exist as anything more, *or anything less*, than complex mixtures of cultured microorganisms. To see this we will visit extreme environments, including hot springs and equally challenging locations such as our own allegedly disinfected homes. I close the book with a vision of life on earth that calls for a reinvention of education in the biological sciences. Lofty goals for a little book, but, as I believe, the miniature is everything.

* * *

Note: each chapter begins with an epigraph from John Milton's *Paradise Lost*. Blind and suffering from gout, heartbroken by the death of his second wife, Milton dictated this epic poem to secretaries and friends between 1658 and 1663. Science intrudes upon the supernatural throughout the work: Milton claimed to have met Galileo in 1638 or 1639 and astronomical revelations feature in many scenes. *Paradise Lost* was published in 1667, two years after Hooke's *Micrographia* and 20 years before Newton's *Principia*. Charles Darwin carried a miniature edition of the poem on H.M.S. *Beagle* in the 1830s and committed large sections to memory.

Milton's advocacy for the power of God as the "high creator" jars with the Darwinian vision of evolution. Yet the poet's telling of the wonder of life was reflected in Darwin's passion when he wrote that his mind was "a chaos of delight" traveling in the tropics. Growing consciousness of the eons of time involved in the creation of the species in the young scientist's collections added to his sense of awe. Milton and Darwin were at the same thing. The Victorian naturalist was on the right track, but we forget how far we have to go today in exploring our planet's riches. Milton isn't a scientific guide, but he may help us revel in the experience. I did not plan how to end this book when I began writing, but the lines from Milton that open the first chapter settled months later on the conclusion of the project in a way that surprises me still.

1

Eden

And all amid them stood the Tree of Life,
High eminent, blooming Ambrosial Fruit
Of vegetable Gold;
—Milton, *Paradise Lost*, Book IV

I can see down to the pond from my second-floor study, its mirror surface shimmering from the impact of raindrops. The goldfish are too deep to be visible from up here, but I imagine them waving their tails slowly at the bottom, just above the silt. Adam, my stepson, and I dug a curvy crater a decade ago, dropped a molded polyethylene liner into the hole, filled it with water, and added a quartet of fish. Following suburban etiquette, we arranged slabs of local limestone around the edge to hide the plastic rim. A conscious act of creation with satisfying results: most of the tree of life situated itself in the water that year. No Leviathan, but the rest is there. It is December now, and this Eden is subdued; fish sleepy, frogs hibernating, and insects, worms, protists, fungi, plants, and bacteria chilled in their multifarious hibernations. The rain keeps pouring this morning and the tree of life is safe, above and beneath the centrifugal ripples.

Even for secularists, the Biblical Garden of Eden is a powerful symbol for all that we have lost. When we assess the qualifications of scraps of the biosphere for special attention, Eden is often used to describe lush forests, grasslands supporting great herds of mammals, or green sea turtles gliding over coral reefs. "Team discovers lost Eden amid forgotten forest in Mozambique," was a news headline a few years ago, reporting that researchers had identified new species of butterflies, reptiles, and populations of birds in a mountainous site first spied with Google Earth. Stories like this are unexpected gifts in the twenty-first century, offering fleeting solace for those of us troubled by the shadow thrown over the rest of life by seven billion apes multiplying toward certain oblivion. Putting the annihilation of our species aside for now, sharing information on forgotten forests is imperative, but by doing so we risk sidestepping a wider truth. Every one of us resides in Eden, we never left: "And all amid them stood the Tree of Life."

Let's return to my pond. My property, like the rest of the planet, is five billion years old in round numbers. When it was half a billion years old it was 20 degrees south of the equator and submerged by a warm, shallow sea. Ohio was in the Caribbean phase of its climatic history.[1] Corals grew on the seabed, mixed with sponges, crinoids (sea lilies), and thickets of branched bryozoans. Shellfish were abundant, including brachiopods and other bivalves that would not seem out of place resting on the crushed ice of a raw bar today. The oysters of the Ordovician were prey for starfish and trilobites that browsed over the shell bed. Jellyfish and conical nautiloids swam in the clear water, and top predator status was held by eurypterids, or sea scorpions, four-eyed relatives of horseshoe crabs.

This marine enterprise was powered by plankton for millions of years in the Late Ordovician, and the remains of this ancient aquarium—the downpour of shells and skeletons and scrubbings from the seabed—formulated an 800-foot thick layer cake of limestone and shale in southwestern Ohio. Every fragment of the slabs around my pond was once part of a living thing: the slipped discs of crinoid stems, half shells and whole shells and broken shells, bryozoan tubes, horn corals, and, throughout, brown flakes of crushed trilobites.

There is no trace today of the 440 million years that elapsed after the Ordovician because all the sedimentary evidence was scraped away by glaciers, the last of which towered more than 1,000 feet above my yard—twice the height of the Washington Monument—and receded 14,000 years ago. The ice sheet voided deep deposits of clay, sand, and gravel and this, in time, became blanketed with hardwood forests that survived until the arrival of my European forebears.

Then, beginning late in the eighteenth century, we set the woods ablaze, turned the enormous beech trees into charcoal for iron smelters, sent the Native Americans packing to Oklahoma, and civilized the land for Presbyterian farmers. Part of the "civilizing" influence meant wholesale decimation of wildlife, including the villainous extirpation of the passenger pigeon. Finally, in the closing years of the twentieth century a developer paid an exhausted dairy farmer and his wife to retire in a Florida condominium complex, cut a road through the land, trucked away the richest topsoil to a garden center, and built my house on the wettest spot. Not content with the natural drainage problems on the property, Adam and I dug our pond to ensure that one spot was wet year round. Billions of years have seen this spectacle distilled from a molten planet that sits 8.3 light minutes from the sun.

* * *

The pond is shaded by a large red mulberry tree, a smaller ash, and an elm. Mulberries grow in abundance in Ohio. Their fruits look a bit like raspberries, but they are close to inedible and I'm far too lazy to make them into wine. They are, in many senses, "trash trees," plants of no commercial value beyond firewood. The biological value of this tree, beyond its participation in the perpetuation of the red mulberry species, lies in the real estate that it offers other organisms. Years ago, this tree split at its base and a gaping wound between the two main branches began weeping brown sap. The wound has never healed because the branches pull apart in the wind, continually reopening the fracture. Despite its injury, this sylvan Fisher King creates a magnificent canopy of leaves every year and shows no bare branches. Birds are hidden from view up in its crown, but the resilience of this dinosaur lineage is clear from the chatter of sparrows, squawking of blue jays, and hammering of woodpeckers. There are some magnificent insects up there too. Cicadas provide the percussion from late summer until early fall. Their voices distinguish four or more species: one sounds like a gravel-filled gourd, others chatter and chirp like birds, one has the high-pitched whir of a jet engine, and still another sounds like a child bouncing on a trampoline with rusty springs. A giant green mantis, the biggest insect I have ever seen, plopped from the canopy one afternoon.

Looking at the mulberry from my desk, I can see its fissured bark and the chocolate-brown ooze from the wound. Not a very interesting thing: just a trash tree. But as one approaches the plant, more and more life becomes visible. The bark is embellished with a gray-green powdering of tiny lichens. The dots are fungal colonies, or thalli, plus the cells of their photosynthetic algal partners. These tiny organisms are transformed by magnification with a hand lens. Each greenish dot is a raft with ragged edges that curl into the air, and some are decorated with the spore-shooting cups of the fungal partner. For much of the year the thalli are bone-dry, but once rejuvenated by rain, the filamentous cells of the fungus resume growth, the algal cells swell and begin producing sugars, and the cups blow smoking rings of spores into the air.

So far, I have identified a few of the most conspicuous pond area residents in this narrative. Mulberries are flowering plants, lichens are fungal sandwiches filled with photosynthetic algae or bacteria, and insects and birds are metazoans, also known as animals. Aristotle and his botanist pupil Theophrastus would have scored poorly had they been tasked with describing the macrobiology of the pond's surroundings, recognizing the animals and the plants, but having no category for the lichen or its component parts. The crude Aristotelian bifurcation of life held sway with minor interruptions, including the genesis of binomial classification by Linnaeus in the eighteenth century, for more than 2,000 years. Protists and bacteria had been observed, of course, soon after the invention of the microscope in the seventeenth century. Two centuries later, biologists Richard Owen, John Hogg, Ernst Haeckel, and others took the first stabs at making sense of the spectacle of life revealed by magnification by creating separate categories for microscopic organisms.[2] And 150 years later, the challenge of developing a harmonious classification of organisms remains.

Cell biological research in the twentieth century made it clear that the diversity of living things was far greater than imagined previously. Early work with the *electron* microscope in the 1950s, coupled with advances in biochemistry and molecular biology, revealed that cells that looked quite similar when viewed with a *light* microscope were constructed according to a wealth of different patterns. The light microscope magnifies things by up to 1,500 times, which is dwarfed by the million-fold or greater enlargement enabled by the electron microscope. Where the light microscope had shown a pollen grain as a spiky ball, the electron microscope revealed a walled plant cell whose surface was decorated with almost impossibly intricate patterning, produced by layers of interlaced polymers. Spots in the cytoplasm of other cells enhanced for light microscopy by staining were resolved clearly as mitochondria with twin sets of membranes. And the more powerful microscope showed that the seemingly clear cytoplasm bathing mitochondria was occupied by a network of interconnecting membranes, vesicles, protein-synthesizing ribosomes, and threaded throughout by a skeleton of filaments.

The electron microscope was a game changer, providing scientists with the information about subtle differences in subcellular structure that allowed them to reimagine the diversity of living things. A pair of cells swimming in a drop of water might have green chloroplasts, yet differ in most facets of cell structure as much as the scientist and the lettuce on her sandwich. New affinity groups for organisms were proposed. Some were sensible, others ludicrous, which reflected the dearth of objective information about evolutionary relationships. The view of the cell provided by the electron microscope recommended varied alliances, but allowed too much subjectivity on the part of the investigator guessing how green blob A might be related to brown blob B.

This began to change as methods for sequencing DNA were developed in the 1970s and the kinds of cells that had been least distinguishable using microscopes were deported to a different biological universe. These were the prokaryotes, the bacteria. Comparisons showed that certain kinds of bacteria housed such different versions of a gene shared by all organisms (which encodes a portion of the ribosome) that they should be placed in a separate domain from the other bacteria and from the eukaryotes. These bacteria were named *archaebacteria*, and later, simply, *archaea*. The second prokaryotic assemblage became the *eubacteria*, or, simply, *bacteria*.[3]

Differences between genes became the reliable metric for understanding the relationships between organisms because, to put it as plainly as possible, evolution *is* genetic divergence. In the last 30 years, this genetic approach has changed biology more than anyone could have guessed, and today's picture of biodiversity presents an irrevocable challenge to many of the core assumptions about life on earth. Which brings me back to my pond. In addition to bacteria and archaea, its waters are home to *amoebozoans, hacrobians, stramenopiles, alveolates, rhizarians, archaeplastids, excavates,* and *opisthokonts.*[4] The eight names for eukaryotes are described as supergroups and you're in good company if you haven't heard of them before. The textbook parceling of eukaryotes into four kingdoms—animals, plants, fungi, and protists— hasn't been a serious reading of the facts for a long time, but it has shown remarkable tenacity. This explains why the majority of professional biologists couldn't tell you what a hacrobian looks like, nor what it feeds upon, or where it lives. But plenty of them would have a few things to say when they learned that the hacrobians include the haptophyte algae. Haptophyte algae are quite important things, playing a major role in the carbon cycle, which is something that we need to be paying attention to as the earth gets warmer and the sea level rises. Humans are members of another of these supergroupings. (Can you guess which one?)

Beyond the challenging names, however, the list is filled with an awful lot of life that almost nobody appreciates and this is a problem. Moreover, the list of supergroups doesn't include the location of a great deal of genetic information. Viruses are the most abundant biological entities on the planet.

* * *

Biologists have found many ways to represent the diversity of organisms and indicate relationships between them. Trees of life, with the tips of branches labeled with the names of groups or occupied by drawings have been popular since the studies of Ernst Haeckel, mentioned a little earlier. Other nineteenth-century biologists, including Darwin of course, had used branching stick diagrams to show hypothetical relationships between organisms, but Haeckel took the tree of life metaphor further and drew gnarled trunks and twisted branches in the 1870s. His trees represented putative evolutionary relationships by organizing related organisms as twigs emerging from the same branch.

Today's diagrams quantify kinship using branch lengths as measures of the similarity between single genes, groups of genes, proteins, or, most reliably, by comparing whole genomes. Trees can be rooted or unrooted. Unrooted trees show the relatedness of organisms without indicating ancestry. Rooted trees converge upon a single node (the root) that represents a common ancestor that probably suffered extinction a long time ago. Organization of the rooted tree is aided by the inclusion of an outgroup, which is a known organism or group of organisms, chosen as a plausibly distant relative that unites everything else in the tree. (Rabbits serve as a distant outgroup to help root a tree of primates.) Trees that are laid out horizontally show close relatives one above the other on adjacent branches. Connections between any pair of organisms, or groups of organisms, are discovered by following the branches back to shared branch points or internal nodes. The passage of time is often represented in these horizontal trees, with the lengthiest pathways to an organism from the ancestor indicating the longest evolutionary journeys. When good fossils are available it is possible to estimate actual time intervals. Unrooted trees lack the implication of time and simply organize the organisms into groups based upon their similarities. Genetic comparisons are de rigueur today, but cell biological, developmental, anatomical, and other characteristics can also inform the relationships in a tree.

When we are looking at the relationships between supergroupings of organisms, like the amoebozoans and hacroblans mentioned above, we are deluged with uncertainties about their evolutionary histories. The fact that there are a number of competing models about the origins of these groups and their interrelationships reflects the extraordinary level of activity in this area of scientific research.

A compelling way to illustrate these fat slices of life is to create a circular diagram and root the groups at the hub of the wheel, which represents their common ancestor. We don't know much about this primordial eukaryote, but the certainty of its existence allows us to show some of the relations between the supergroups. Stramenopiles, alveolates, and rhizarians have enough in genetic common to suggest that they share the same link to the first eukaryote. In other words, these SAR eukaryotes diversified from a common ancestor after this organism had split from the ancestors of the other supergroups. This should be clear from the wheel diagram in Figure 1: the SAR eukaryotes are connected to the hub via a single spoke. With this diagram in hand it's time for an opportunity missed by Jacques Cousteau and his aqua-lunged heirs: Ohio Pond Quest. No submersibles are needed for this adventure, only some bottles and a microscope.

We begin with the amoebozoans. The archetypal amoeba, the exemplar of simple organisms recognized by anyone who has taken middle school biology, is the amoebozoan called *Amoeba proteus* (Figure 2). This species is the best-known microbe, the only manifestation of microscopic life that attracts name recognition across vast divides of culture and educational system. I have little

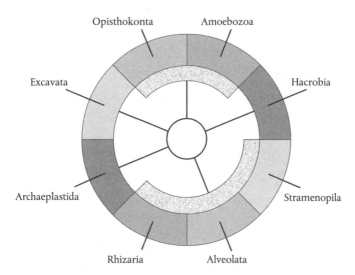

FIGURE 1 Wheel of eukaryote life. Eight supergroups arranged around the circumference are linked by spokes to the central hub symbolizing the ancestral eukaryote derived from prokaryote progenitors. Common prototypes that unite some of the groups are indicated by the intermediate segments.

idea what teenagers are taught about biology in North Korea, but I'm willing to bet that boys and girls in Pyongyang would have something to say about this symbolic blob of cytoplasm. The same is true of Amish offspring schooled in Ohio, though their religious curriculum saddles them with a pre-Darwinian view of life. *Amoeba proteus* spooges itself through the pond silt, redistributing its cytoplasm into limbs called pseudopodia, advancing at one end, retracting at the other. The cell defines a front and a rear end, at least on a temporary basis, leading with its pseudopodia. Motion for *Amoeba* is, most often, a predatory impulse. When successful, the cell embraces its microbial quarry in a pseudopodial hug, casts the morsel into a vacuole, and showers the terrified bacterium with digestive enzymes. This is phagocytosis. Later, any waste materials are voided by the reverse of the feeding process at the posterior end of the amoeba. No mouth, no anus, but the essence of all animal life is there. (And yet some insist, contrary to evidence, that there is something more, a greater essence, to *Homo sapiens*.)

Amoeba puts on a captivating show under the microscope. It is a large cell, sometimes more than half a millimeter in diameter,

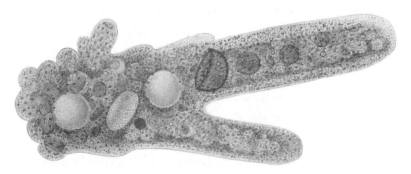

FIGURE 2 *Amoeba proteus* (Amoebozoa).
Source: J. Leidy, *U.S. Geological Survey of the Territories Report* 12, 1–324 (1879).

which is one of the reasons it was adopted as "The Cell" in early biology textbooks (the other reason is that it is easy to collect). There is great smoothness to its locomotion, a continuous spilling of the cytoplasm from one place to another as the cell, the whole organism, tries to escape the bright illumination of the microscope lamp. *Amoeba* is derived from a Greek word meaning change, and the species name refers to the Greek sea god Proteus, one of the deities referred to by Homer as "old man of the sea," capable of transforming his shape at will.

The genetics of amoebozoans are remarkable. *Amoeba proteus* has hundreds of chromosomes, encoding a genome 100 times bigger than ours.[5] The genome of another amoebozoan, called *Polychaos dubium*, may be double the size of the *Amoeba*, ranking as the largest of all information packages in a nucleus. Not all of this information is useful. Genome size offers a poor match to the number of functional genes, and amoebozoans whose genomes have been studied in great detail certainly encode fewer genes than our DNA. The tremendous amount of DNA in these cells is a consequence of the passage of the better part of one billion years of being the same kind of thing: copying of the genes (duplication), copying of the entire set of instructions followed by copying of the copies (polyploidy); redundancy and dysfunction of old genes (genomic fossils); infection by retroviruses and the assimilation of their genomes, and the transmission of genes from other microbes (horizontal gene transfer). At every opportunity, natural selection prunes out the damaging instructions and leaves the non-coding "junk," as well as the functional sequences, to carry themselves into the future. Evolution has been chiseling away at the amoebozoan DNA, but, for obscure reasons, these cells are streamlined in form, but obese in genetics.[6]

In this brief dip in the pond it is impossible to do more than offer snapshots of each group, a few tweets per slice. Reaching beyond my pond, to freshwater habitats throughout southern Ohio, one could spend decades describing new kinds of amoebae and never run out of subjects. Biologists used to do this sort of thing. Joseph Leidy, who authored a lengthy monograph on these organisms published in 1879, wrote, "How can life be tiresome so long as there is still a new rhizopod undescribed?"[7] (I used Leidy's illustration of *Amoeba proteus* in Figure 2.) We have come a long way, and been humbled greatly, since Pliny the Elder embarked upon his hopeless plan to record "all the contents of the entire world" in his encyclopedic *Naturalis Historia*. Modern research on microbial diversity has shifted from the microscope to the automated sequencer. This has revealed an astonishing breadth to the variations in the smallest forms of life. Two thousand years after Pliny, however, we are little closer to completing his catalog. Harvard ethologist E. O. Wilson recommended another shot at this task through his *Encyclopedia of Life* project.[8] The quixotic nature of the endeavor was obvious to anyone who thought about it for a few minutes, and the stocktaking has failed to overcome Wilson's preoccupation with animals.

Another way to consider the diversity of amoebozoans is to think about the range of behavioral sophistication. There is more to the group than squirming around and eating bacteria. *Amoeba proteus* is a solitary protist that manages to send its enormous genome down the river of time without mating. Though it may have evolved from ancestors that comingled their genomes from infrequent time to time, there is no evidence that it partners with other amoebae today. It chases its prey, feeds, and digests, divides by mitosis and yields a pair of cells that repeat the cycle. Other amoebae are highly social. Slime molds are examples of social amoebae and dictyostelid slime molds are the best studied.

The life cycle of *Dictyostelium discoideum* is one of the basic processes taught in introductory biology classes. Its amoebae act as solitary cells when there is plenty of food around, but when the bacterial feast thins, the amoebae stream together to produce a tiny slug, or grex. This grex operates like a multicellular organism, moving as a discrete entity motorized by the migration of 10,000 to 2 million aggregated amoebae within a slimy matrix. After relocating the colony a few millimeters to escape the soil and reach an exposed surface, the grex halts, and transforms itself into a stalk topped by a bulb. The amoebae in the bulb are polymer coated for resistance, and dislodged and dispersed by invertebrates that brush over this slime mold fruit. The genetic signature of dictyostelid slime molds is found in freshwater, but it isn't clear what the social amoebae are doing in a pond like mine. It is possible that they can grex over the goo that coagulates above the water line.

We could stay with amoebozoans for much longer, but so many riches await us as we blow the silt from our nostrils and dip our heads into the water again. Hacrobians in my pond include the cryptomonad alga, *Cryptomonas ovata*. This is one of the many kinds of swimming cells in the ecosystem, a fast-moving organism that propels itself through the water using a pair of hairy flagella. *Cryptomonas* is most active during colder months and seems to thrive beneath the ice in winter. Thoreau wasn't thinking about algae at the time, but his contemplation of life beneath the ice of *his* iced pond is among the most beautiful passages in *Walden*:

kneeling to drink, I look down into the quiet parlor of the fishes, pervaded by a softened light as through a window of ground glass, with its bright sanded floor the same as in summer; there a perennial waveless serenity reigns as in the amber twilight sky, corresponding to the cool and even temperament of the inhabitants. Heaven is under our feet as well as over our heads.[9]

Cryptomonas is a Russian nesting doll of a cell, a composite of multiple organisms that fused during its lengthy evolutionary past. Clues to this are found in the unusual structure of the cell and proof in its multiple genomes (Figure 3). The large nucleus is situated toward the bottom of the cell and is what qualifies the alga as eukaryotic. The nucleus contains chromosomes and these constitute genome I. Genome II is much smaller and resides in the algal mitochondria. The mitochondrial genome encodes less than 50 genes, and these are arranged as a hoop of DNA with none of the intervening sequences called introns that are found in the adjacent nuclear genome. The size of this circular chromosome and the absence of introns are among the

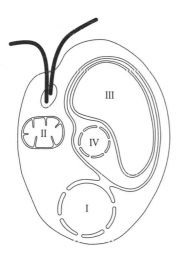

FIGURE 3 *Cryptomonas* (Hacrobia), an example of a eukaryote generated by secondary endosymbiotic fusion of an amoeboid cell with a photosynthetic red alga. Evidence of this chimeric origin is found in the presence of four membranes surrounding the chloroplast and in a structure called the nucleomorph trapped between the membranes. The nucleomorph is an accessory genome derived from the nucleus of the red algal partner.

clues to the bacterial origin of the mitochondrion. Much the same can be said of our cells.

But there are a lot more layers to the doll that is *Cryptomonas*. Like the mitochondrion, the chloroplast has an irrefutably bacterial origin and contains a bacterial chromosome in its center. This is genome III. Plant cells have the same photosynthetic organelles and so they share the three-genome complexity of these algae. *Cryptomonas* has another genome—genome IV— trapped between the multiple membranes that wrap around its chloroplast. Genome IV nestles within its own set of membranes and is called the nucleomorph. It is a miniaturized nucleus that contains three chromosomes encoding 500 or so genes.

This chimera is a particularly complex example of endosymbiosis, though we'll encounter one higher level of complexity in a few pages. The membranes that surround mitochondria and chloroplasts were derived from the membranes of the arrested organisms and from the vacuoles formed in the capture process. The capture method was phagocytosis, the feeding mechanism of the amoebozoans. Feeding that led to the creation of organelles must have stopped after the gulp. If digestion had followed there would be no mitochondria or chloroplasts. Analysis of the genes in mitochondria and chloroplasts has allowed molecular biologists to make reasonable guesses about the types of bacteria that survived digestion to be transformed into today's mitochondria and chloroplasts. *Cryptomonas* adds a level of complexity because it seems that its chloroplast was derived from the absorption of a red alga rather than a bacterium. Evidence for this is found in the red-algal instructions found in that nucleomorph, and the fact that the *Cryptomonas* chloroplast is surrounded by four membranes rather than the usual pair in plants.

In addition to its red-algal nucleomorph, *Cryptomonas* cells have another remarkable structure called the ejectisome. Ejectisomes are coiled ribbons of protein situated just beneath the surface of the cell. The largest of these point into the groove in the cell surface from which the paired flagella arise. If the cell is irritated, the ejectisomes discharge at high speed, the coils slipping over themselves, each snapped like a roll of Christmas wrapping paper into a lance. An eloquent blogger and "aspiring protistologist" with the pseudonym Psi Wavefunction called cryptomonads "solar-powered armored battleships."[10] The natural function of the ejectisomes isn't settled. They may operate as defenses against predators that browse on algae in the pond, pushing the cell away from the spiky appendages of water fleas and other hazards. Beyond the pond, marine hacrobians include the haptophyte algae that I mentioned at the beginning of this suburban safari.

The third group of eukaryotes, the stramenopiles, is well represented in my pond. Some can be seen without a microscope. These are the white colonies of water molds that spread around the floating carcasses of insects and become tipped with sporangia that liberate swimming spores into the water.[11] The spores of water molds are called zoospores. They have paired flagella, but only one of each pair is covered with hairs, unlike the doubly hirsute swimming cells of the alga *Cryptomonas*. After a few minutes or hours of swimming, the zoospores stick to a surface and encapsulate themselves in a wall to form a cyst. Water molds can use encystment as an entry to a vegetative state when they shut down for wintry weeks or months, taking a break from their otherwise endless quest for food. At other times they swim, encyst, and then emerge from the cyst and swim some more, wagering that brief bursts of high-speed locomotion are more effective at locating food scraps than marathon swimming until metabolic exhaustion. This makes perfect sense for something that lives in a pond waiting for an insect to misplace a hoof and fall from the overhanging trees.

Most of the water molds are saprotrophs that feed on the remains of animals and plants using the same absorptive feeding mechanism as fungi. Less is known about the ecology of the predatory stramenopiles that infect rotifers and nematode worms, but researchers have celebrated the awesome weaponry utilized by species of *Haptoglossa*. The predatory zoospores of *Haptoglossa* behave in a very different way than saprotroph spores. *Haptoglossa* may hibernate too, but cyst formation also serves as an opportunity to reorganize the cell from its mobile phase into an offensive weapon. In this case, the cyst bulges on one side, forming a new compartment shaped like a gourd into which the water mold transfers all of its cytoplasm. In the next four to five hours the cell contents are reorganized to create a marvel of microbial artillery called the gun cell (Figure 4).[12] This includes a dart whose tip is surrounded by a set of overlapping cones that hold it in the center of a barrel. The dart sits at the end of a tube that is coiled in the cell behind the gun barrel. When a nematode brushes over the tip, or beak, of the gun cell it becomes stuck by a pad of adhesive. As it struggles the gun cell fires, pushing the dart through the nematode cuticle and uncoiling the tube behind. The tube rolls inside out as it follows the dart and transfers the gun cell contents into the doomed animal. (Picture a latex glove with a finger pushed inwards, then grasp the end of the glove in your imagination and blow it out to extend the finger, and think about that finger representing the tube traveling into an agonized worm, and you will either be very confused, or closer to

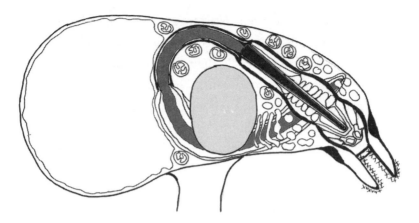

FIGURE 4 *Haptoglossa* gun cell (Stramenopila).
Image courtesy Gordon Beakes, Newcastle University.

imagining how this works. The worm would need to be as big as a school bus to model this to scale.) Darting and transfer occur in a few tenths of one second and then the worm pulls itself free and glides on its way, accompanied by its lively parasite.

The water mold grows inside the worm, digesting the animal's tissues to fuel the ovoid bodies, or thalli, which constitute the feeding phase of *Haptoglossa*. When the worm is dead, or dying, these thalli convert themselves into sporangia that expel a new generation of zoospores into the water, repeating the cycle that carries the water mold's genes into the future. Some *Haptoglossa* species attack rotifers rather than nematodes, and there is no shortage of either prey in the pond.

Diatoms are stramenopiles too and there are lots of these gliding around in the silt at the bottom of the pond, up the mucilage-coated sides of the pond, and over the submerged leaves of dangling plants (Figure 5). Diatoms in pond water are often viewed in biology classes when students are taught how to use microscopes. Beyond their assistance in adjusting the illumination, diatoms and other single-celled organisms can be a source of fascination for students if they are encouraged beyond the cynicism common to their youth. This may mean an instructor willing to leap around and energize the room of teenagers by pure force of will. The diatoms in ponds are shaped like canoes and glide between obstacles on the microscope slide, nudging

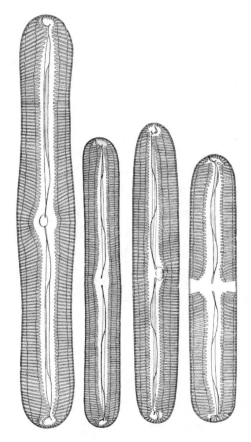

FIGURE 5 Freshwater species of the diatom *Pinnularia* (Stramenopila).
Source: F. Hustedt, *Bacillariophyta (Diatomere)* (Jena: Gustav Fischer, 1930).

other small organisms aside, hitting larger obstacles, and revers-ing direction. They look a lot like claymation characters in those jerky stop motion animations when the filmmaker limits the number of movements to retain the staccato form. Diatoms move smoothly during periods of motion between obstacles, stopping, turning, and going again when they hit things. This movement, called gliding motility, seems to involve the extrusion of muci-lage, but we don't know very much about how it works.

The enigmatic nature of gliding motility—which may be a eu-phemism for the deficiencies of the biologists who have studied this problem—is a useful thing to point out to students: scien-tists, albeit not many of them, have been looking at diatoms for more than 200 years, but have failed, so far, to offer a persuasive explanation for their mobility.

The diatom shell, or frustule, is a glass box patterned and per-forated with tiny holes whose arrangement is an aid to identifying the 10,000 described species. Genetic evidence shows that dia-toms are related to some of the tiniest planktonic organisms called bolidophytes and, more distantly, to giant kelps and other kinds of brown algae. All of these organisms are photosynthetic. Their chloroplasts seem to have been derived from ancient endo-symbiotic congress with a red alga. This happened such a long time ago that there is no trace of the original algal nucleus, the nucleomorph found in the cryptomonads. I'll return to diatoms later in the book because they are one of the groups of micro-scopic things that have shaped the chemistry of the biosphere for a little bit longer than it has been troubled by the apes.

Alveolates are also well represented in the pond. Dinophysoids, also known as dinoflagellates, call attention to themselves in every drop of water by the helical path and speed of their swimming.

They are the fastest swimmers in the pond, and *Peridinium* is one of the commonest. It has a rigid armor which is assembled as a series of cellulosic plates that fit together *beneath* the cell's membrane rather than on the outside like a diatom (Figure 6). The subsurface arrangement seems counterintuitive for protection, given the vulnerability of the cell membrane to puncture. It may help visualize the dinoflagellate structure by considering the way that the bones that form our skulls sit under the skin. The plates form a pair of grooves: one that runs around the belly of the cell, the other running down half the cell at a right angle to this. Each groove guides a flagellum. The one around the middle is coiled tightly within its groove (picture yourself wearing a spring as a belt). This vibrates in a way that causes the cell to pirouette as it

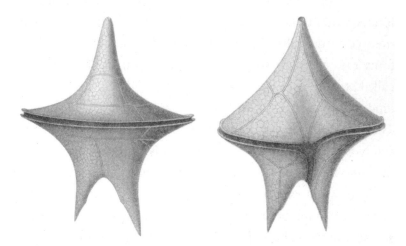

FIGURE 6 Two sides of the marine dinoflagellate *Peridium grande* (Alveolata). The grooves in the cell surface serve as guides for a pair of flagella that emerge from the junction shown in the image on the right. Common freshwater dinoflagellates have a very similar cell structure.

Source: G. Karsten, *Das Indische Phytoplankton* (Jena: Gustav Fischer, 1907).

pushes through the water. The second flagellum is an accessory motor and also serves as a rudder, undulating from its tip toward its base, providing forward thrust and also causing the rotating cell to fly through a gyre.[13]

Some dinoflagellates are photosynthetic, others lack chloroplasts and feed on diatoms and other planktonic microbes. The biggest marine dinoflagellates feed on larger prey including copepods. My pond *Peridinium* is photosynthetic, and the origin of its chloroplasts takes us back to the nesting doll analogy. Dinoflagellate genetics are even more complicated than the cryptomonad algae because some of their chloroplasts *are* cryptomonad algae. There are multiple possibilities for doll assembly, because dinoflagellates appear to have eaten other kinds of photosynthetic protists including green algae, diatoms, and haptophytes, each time creating a novel compilation of genes. The collaborating genomes are streamlined as millions of years pass, with the deletion of redundant genes and the transfer of others between compartments obfuscating these ancient mergers. The molecular phylogenetic research that has picked apart these genomic fossils has been one of the great success stories of biological research in recent decades.

The rhizarians complete the SAR triad of eukaryotes. Marine ecosystems harbor much of the diversity within this supergroup, including the breathtakingly beautiful cells of the planktonic radiolarians and foraminiferans. Rhizarians are represented in the freshwater of my pond in the shape of euglyphids. (I'm sorry for all the names, but there is no other way to survey life without carving it up and pinning nametags to chunks of convenient size.) Euglyphids, like many of these groups, lack a common name. "Testate amoebae" may be the closest they have to a label that most biologists recognize, but, to confuse things, the euglyphids aren't the only amoebae that form "tests" or shells. Some amoebozoans shell themselves too. The euglyphid shell is produced as a series of plates or scales within the cytoplasm (Figure 7). These are secreted by the amoeba and assembled on its surface, leaving an opening at one end. They are shaped like Greek amphorae, without handles. The shell affords protection and the amoeba

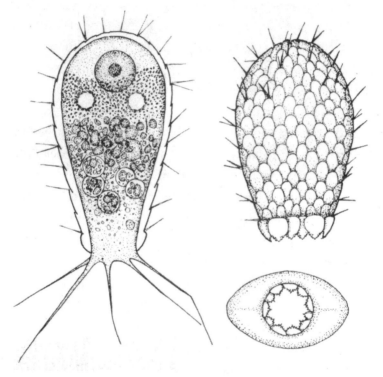

FIGURE 7 Testate amoeba, *Euglypha strigosa* (Rhizaria). Thread-like (or filose) pseudopodia emerge through the ring of teeth surrounding the open end of the shell. The amoeba glides over surfaces holding its shell in an upright posture.

Source: G. Lüftenegger et al. *Archiv fur Protistenkunde* 136, 153–89 (1988).

extends itself through the opening in the form of thin, highly mobile filaments. These operate like pseudopodia, moving the cell and its jar over surfaces in the pond and trapping bacteria. They remind me of hermit crabs. Euglyphid shells are manufactured from silica and form pretty microfossils, revealing that these microbes have existed in an array of shapes like those found today for more than 750 million years.

Archaeplastids include the plants, but there is a lot more to this supergroup than the wounded mulberry that shadows the pond water. Beneath the mulberry its green algal ancestors soak and swim. These are the species whose genes were carried aloft when plant life on land figured out how to cope with massive doses of mutagenic radiation and elevated their chloroplasts toward the sun. The green alga *Spirogyra* is related, very distantly, to the kinds of protists from which the plants evolved. Like *Amoeba*, *Spirogyra* is among the handful of microbes showcased in many biology textbooks (Figure 8). Its sexual behavior is a major reason for this top billing. The alga grows in the form of filaments divided into short cylindrical segments by cross walls. Each segment is a cell occupied by a single prominent nucleus. The filaments are fueled by photosynthesis and *Spirogyra* chloroplasts are big and beautiful, bright green helical ribbons that wind around the inside of the cell. The ribbons have studs embedded on their surface called pyrenoids, in which the glucose synthesized through photosynthesis is stored in polymeric form as starch. Sex takes place when adjacent filaments produce short tubes that fuse at their tips to create a fluid continuum between pairs of partnering cells. One cell becomes the giver, the other the receiver; the giver's contents move in an amoeboid fashion through the tube to merge with the receiver.

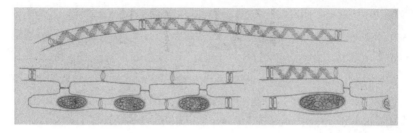

FIGURE 8 *Spirogyra* (Archaeplastida). Single filament with spiral chloro-plast (top); conjugating filaments with zygotes in lower cells (bottom).
Source: W. F. R. Suringar, *Dissertatio Botanica Inauguralis Continens Observationes Phycologicas in Floram Batavam* (Leovardiae, 1857).

The textbooks refer to the giver as the male, but there is nothing to recommend this terminology from a genetic standpoint. Sex is between consenting androgynous filaments, and Leviticus offers no opinion on this outrageous behavior. The comingled cyto-plasm forms a wall and the resulting nugget can serve as a resting spore, allowing the alga to survive the colder months before ger-minating into a fresh filament. *Spirogyra* doesn't produce any swimming cells, but plenty of green algae do attract the micros-copist's attention by flashing across the field of view. The diversity of the green algae in fresh and saltwater habitats is quite aston-ishing. They range from tiny floating spheres and stars, to the beautiful multicelled motile colonies of *Volvox*, to leafy seaweed, to rigid platelets, to the siphonaceous greens whose enormous cells are controlled by thousands of nuclei. And all the green algae occupy one sliver of the archaeplastid eighth of the circumference of the wheel of life.

One thing I must point out is that the organisms that I'm de-scribing are active in the pond at different times of the year. Air temperature fluctuates from an average winter low of minus 6 °C

(21 °F) to a summer high of 31 °C (88 °F). Populations of microbes rise and fall with the seasons. When the water warms in spring, the leafing mulberry throws its shadow and the fish escape the hottest afternoons by settling in the cooler water at the bottom of the pond. As the days shorten, falling leaves fertilize the water and the pond cools, freezes, then thaws again. The water is almost transparent in the early weeks of the summer when it clarifies for a month or two after the previous year's mulberry leaves have thoroughly fragmented and formed the rich finger-deep sediment that sustains so much of the life in the ecosystem. On bright days, the water is lit by shafts of sunlight that stream through openings in the leaf canopy and sparkle all the way down to the fine-tilled silt. Appearances are often deceptive. The water is certainly cleaner than the stuff that flows along the Ganges, but I have no doubt that a glass or two of "L'eau d'étang stagnant de Nik Monet" would cause people accustomed to bottled water to eject their intestines into their underwear.

As the summer temperature rises, the warming water absorbs less oxygen and the pond darkens with new forms of life; biology clouds every drop and one of the loveliest algae appears. This is called *Phacus* (fay-cus) and it's a euglenoid, representative of the seventh supergroup, the excavates (Figure 9). Imagine a transparent heart-shaped box of Valentine's day chocolates, now shrink this so that 50 of them could fit side by side within the space of a millimeter. *Phacus* cells are tiny heart-shaped boxes. The alga is thin and flat, and rotates as it swims. Its transparent cover is made from strips of flexible protein which are assembled as a series of interlocking bands. In terms of its construction, this wall or pellicle is a microscopic version of the vinyl siding on an American house. Beneath the pellicle sit the organelles (candies in our confectionary model). In addition to the usual nucleus, sinewy and bubbly membranes, and mitochondria found in every other eukaryote cell, there are chloroplasts (the confectionary model is making more sense now), and a bright red spot that functions as the *Phacus* eye. This alga has a pair of flagella, though only one of them projects beyond a little pocket at one end of the cell and drives the euglenoid through the water. Trapped in a film of water between a glass microscope slide and a coverglass, the cell seems to change shape, but this is an illusion caused by its strange profile and revolving motion: heart, rod, heart, rod. Like other algae, *Phacus* is a primary producer, one of the organisms responsible for sustaining the non-photosynthetic life forms in the pond.[14] It has an apple-green color, but it isn't a green alga in the sense of being an archaeplastid.

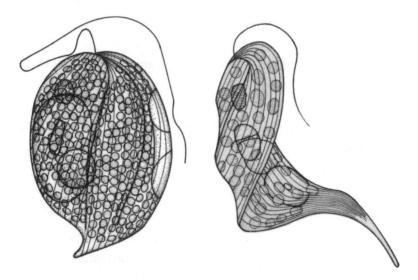

FIGURE 9 Freshwater euglenoid algae, *Phacus platyaulax* and *Phacus raciborskii* (Excavata).

Source: M. J. Perieira and U. M. M. Azeiteiro, *Acta Oecologica* 24, S33–S48 (2003).

The pond contains one more group, or supergroup, of eukaryotes: the opisthokonts. I'm not providing an illustration for this part of the wheel. If you want one you can go look in the mirror. The opisthokonts house the animals, fungi, and nonconformists like the Mesomycetozoea, which parasitize fish and sit somewhere, in evolutionary history, between the animals and fungi. The pond is colonized with lots of aquatic fungi with easily misspelled names: the Chytridiomycota, or chytrids, and the Blastocladiomycota. These are spectacularly diverse, both in terms of their genetic spread and variety of structure. Many grow as inconspicuous globules outside or inside the cells of various algae and other pond life. The ones that grow on the surface of algal filaments penetrate their hosts via a network of fine filaments, like a tiny root system; others produce spectacular branching thalli tipped with reproductive cells that release sperm and eggs. All of these fungi are swimmers, producing zoospores with single flagellar motors that push them through the water.

The single-tailed cell is the signature of all of the opisthokonts. We have a slightly more complex anatomy than the fungi that swim with my goldfish, but our affiliation with these microbes may be read in our genes and is probably reflected in our motile sperm cells, and in the ciliated cells of our airways, brain ventricles, and fallopian tubes.

This consideration of cell structure in complicated opisthokonts brings us full circle in the wheel of eukaryote life. Pond Quest began with the amoebozoans, and there are many kinds of cells in our bodies that operate as amoebae. These include the neutrophils and macrophages of the immune system that consume bacteria using the phagocytotic mechanism of the pond *Amoeba*. The seventeenth-century microscopists recognized the cellularity of every organism, but couldn't triumph over the idea of human exceptionalism. Four centuries later and we remain innocent of the philosophical implications of the amoeba in the room.

Human DNA is likely to be floating around in the pond. The pond pump needs cleaning every week or so and I leave my genes all over the device as I scrape the algal coating from the inlet grill. The range of other animals is fabulous and would require its own chapter for a proper inventory. Two or three frogs engage in their amphibious activities, worms abound, plus rotifers, water bears (tardigrades), *Daphnia* and copepod crustaceans, *Hydra*, insect larvae, and water snails. It seems odd to have reached the end of this survey of eukaryotes with little more than one sentence on animal diversity. But to dedicate more space to frogs, for example, doesn't make sense when we are engaged in an objective survey of life in the pond. Depending upon the chosen scale, or magnification, of the analysis—the threshold of genetic difference between groups of organisms that demands their separation on individual spokes—the animals may be separated from the fungi (more spokes), or held with them in the opisthokonts (fewer spokes). Groups of organisms bleed into one another through their ancestral roots.

An objective survey of life reduces the significance of most of the familiar things that fill a biology course. Animals and plants are slivers of biological diversity. This is obvious when we look at the eight supergroups of eukaryotes, yet we have seen only the backwater of existence by concentrating this description of life upon the eukaryotes. Most of life, most of the biological information in the pond and elsewhere in the biosphere, reproduces itself outside the eight supergroups of eukaryotes. Prokaryotic bacteria and archaea outnumber the eukaryotes by many orders of magnitude. The brightly illuminated circle of pond water on a microscope slide with ten protists contains 10,000 or more bacteria. The prokaryotes lack the exhibitionism of the dinoflagellate or the euglenoid, most of them fill the microscope field of view as a kind of granular backwash. They are like the Milky Way for an astronomer whose primary concern is the Solar System. Only at the highest magnification does this cloudiness resolve as myriad rods and spheres, specks of life jiggling in the water.

There are many more bacteria than eukaryotes, but even more information exists in the form of viruses, which are totally invisible under the light microscope. And beyond the prevalence of viruses, many of the genes in the rest of life in the pond may have originated as viral instructions. The truth of life looks nothing like the traditional tree where almost everything is an animal, and the vastly more numerous oddities are squashed at the base under an icon that looks like an amoeba or a hairy bacterium. To teach a course on biological diversity lasting the standard three-month American semester and to spend more than one class on the animals is to encourage an absurd caricature of life. The way that we teach biology is no more sensible than evaluating all of English Literature by reading nothing but a Harry Potter book.

2

Lenses

that I may see and tell of things invisible to mortal sight
—Milton, *Paradise Lost, Book III*

W hen Galileo was questioned by the Roman Inquisition, the Church had hoped to silence the astronomical truth of the cosmic inconsequence of earth. The cardinals had no idea that the Tuscan genius had committed a far greater sin by modifying his telescope to enlarge the hairs on the back of a fly. Galileo's adventure in biology would prove him a wiser God than the bearded phantom of Catholicism. Through the simple act of magnification, he founded a new method of inquisition, one that would show man's insignificance on his little planet.

The effects of natural lenses were probably recognized by our hominid ancestors. The magnified death throes of an ant drowning in a raindrop on a leaf would have caused the more curious of these apes, crouching under a tree until the storm passed, to grunt and point. Assyrian craftsmen may have been the first to grind lenses from quartz crystals. These were used as magnifiers, or as "burning glasses" to start fires. The earliest spectacles were perched on Italian noses in the thirteenth century, and the development of these marvels has been linked to the translation of an eleventh-century Arabian work by Ibn al-Haytham (Alhazen).[1] Alhazen's *Book of Optics* made a persuasive case for vision depending upon the entry of light into the eye, rather than the Ptolemaic boo-boo of light emission *by* the eye.

The first microscopes may have been produced in Holland in the 1590s by Zacharius Janssen. Caution about the date is necessary because Zacharius was only 10 years old at the time of the earliest claim for his invention. A slightly less youthful Zacharius may have been assisted by his father, but the historical records that might have settled this matter were destroyed in the Second World War. But Janssen, or the Janssens, did make microscopes at some point and the available evidence—contradictory descriptions and a surviving instrument of uncertain authenticity— suggests that these were compound microscopes made by aligning a biconvex eyepiece lens with a second plano-convex lens (one side flat, one curved outwards) held inside a pair of tubes. The tubes were inserted into a sleeve and could be pushed toward one another, or pulled apart, to focus the specimen. This handheld design would have provided the viewer with up to nine fold magnification, making a fruit fly as big as a fingernail, or revealing the surprising bluntness of a sewing needle. The younger Janssen also claimed to have invented the telescope, but this was contested by contemporary Dutch investigators who submitted patents for instruments in the early seventeenth century. The case for Janssen's priority for either instrument isn't strengthened by the fact that he was tried on a number of occasions for the capital crime of counterfeiting coins.

While the Dutch squabbled about priority, Galileo was adapting his telescope design into a microscope. Marveling at the magnification of insects achieved with an elongated version of his telescope, he reworked the optics to produce a tabletop microscope, or *occhiolino*.[2] Galileo's microscope held a pair of lenses and was supported in an upright orientation in a metal stand. The specimen was viewed on a tabletop and magnified up to 30 times, "so that one sees a fly as large as a hen."[3] If we discount Janssen's claims, Galileo must be credited with the invention of the microscope in 1610.[4]

The optical quality of the early compound microscopes was pitiful. Bubbles and flecks in lenses obscured parts of the field of view, flaws in lens curvature distorted the images, and colored halos surrounded every magnified object. Imperfections aside— in the land of the blind, after all, the one-eyed man is king—the Galilean microscope was a scientific wonder. A decade before his infamous trial in Rome, Galileo showed Giovanni Faber, Curator of the Vatican Garden, how to look at a fly with his microscope. In a letter to Prince Federico Cesi, Faber said that he had told Galileo that he was "another Creator, since he makes appear things that no one knew had been created until now." Cesi was a fan of Galileo. He had founded a scientific society called the Accademia dei Lincei (the Academy of the Lynx-eyed), in 1603, more than a half century before the first meetings of the Royal Society of London. Galileo joined the society in 1611 and gave Cesi one of his microscopes. Faber, who became the Lynx's secretary, is credited with naming both of Galileo's revolutionary instruments: the telescope in 1611 and microscope in 1625.

Galileo's dangerous ideas were supported by his friends in the Lynx. The society published his *Il Saggiatore* (*The Assayer*) in 1624, which was an ill-founded critique of the work of a Jesuit astronomer on comets. The Linceans followed with their own three-part thesis on bees, which included a sheet of engravings of the insects whole and a collection of their disarticulated parts.[5] Galileo's microscope was used in these investigations, and the bees were the first printed illustrations showing microscopic details. *Il Saggiatore* and the bee studies paid homage to Pope Urban VIII, Maffeo Barberini, and reproduced his family crest, which was a trio of bees. The exquisite figures were based on the observations of fellow Lyncean, Francesco Stelluti, and the society anticipated that these works would curry favor for themselves and for Galileo. A manuscript that evoked the pastoral pleasures of beekeeping and celebrated the insects cherished by the Pope promised welcome distraction from the considerable administrative burdens of being God's chosen representative. Pope Urban appears to have been happy enough with this obeisance. But the bees were subversive too. Stelluti's illustrations showed the multi-paned eyes of the insects looking like tiny Tudor windows, the hairiness of the legs, the intricacies of the maxilla, labium, and nectar-slurping tongue, and the sharpness of the sting. Nobody had seen bees like this. Hidden wonders of creation—of God's creation—were made visible with Galileo's microscope and the investigations of the Linceans. In *Il Saggiatore*, Galileo spoke of mathematics as the language of God and an essential tool for escaping the "dark labyrinth" of ignorance. Less abstract than algebra, magnified bees hinted at the rising power of unfettered scientific investigation.

For reasons obscured by the passage of 400 years, Galileo chose to turn his tolerant Pope into an enemy in an act of almost suicidal bravado. *Dialogue Concerning the Chief Two World Systems*, published in 1632, was a heretical book that made Galileo's case for the Copernican vision of the heliocentric Solar System. The *Dialogue* used the inflammatory device of an argument between three characters: a Copernican advocate, a neutral voice, and a character called Simplicio who articulated the opinions of Galileo's ecclesiastical critics.[6] Simplicio was named after Simplicius of Cilicia, an admirer of Aristotle, but could not avoid being read also as a fool. During an earlier investigation of Galileo's work, Cardinal Robert Bellarmine said, "To assert that the earth revolves around the sun is as erroneous as to claim that Jesus was not born of a virgin." The subsequent publication of the *Dialogue* was a deliberate taunt against the Church. The 68-year-old scientist was summoned by the Inquisition, found "vehemently suspect of heresy," and spent the remaining decade of his life under house arrest in Tuscany. (Which reminds me of the Monty Python sketch, *The Spanish Inquisition,* in which "the comfy chair" is among the dread instruments of torture.) Galileo's cosmology was an overt threat toward the Catholic Church. *Dialogue Concerning the Chief Two World Systems* was placed on the Vatican's *Index of Forbidden Books,* where it remained for 200 years. The significance of the Linceans' exploration of bee anatomy was ignored by the Inquisition.

The illustrations in Robert Hooke's *Micrographia,* published in 1665, revealed the considerable advances made in microscope design in the following half century.[7] Hooke's microscope was more complex than its predecessors, with a series of three lenses within four concentric tubes. He used an oil lamp as a more reliable light source than sunlight, and focused its flame onto the specimen using a water-filled globe and "bulls-eye" condenser lens. It was a beautiful instrument, finished with a gold-tooled leather wrap, but was difficult to use and suffered from the optical handicaps of all of the early compound microscopes.

Perusing Hooke's plates one has no inkling of the struggles that he endured in his exploration of the intricacies of insects, slivers of cork, and fungi. A two-page spread displays the head of a hoverfly far wider than the face of the reader. The illustration demands one's attention because Hooke's insect *is* the viewer. The huge hemispherical eyes of the hoverfly resemble the honey-combs in beehives; Hooke described their tessellated surfaces as clusters of pearls. The detail goes way beyond the insects in Stelluti's images. Other plates were fold-out displays, including the famous flea depicted as a hairy-legged monster as big as a cat, and a monstrous head louse gripping a hair fiber thick as a finger. These plates are grand gestures by Hooke, flamboyant displays of the power of the virtuoso scientist and his microscope. Hooke is presenting a new view of life and of new forms of life: "we now behold as great a variety of Creatures, as we were able before to reckon up in the whole *Universe* itself." He describes this revelation in the preface to *Micrographia* as "a universal cure of the mind."

Few of Hooke's readers may have digested the lengthy text, but the plates were mesmerizing and the book was a best seller in the plague year before London's Great Fire. Samuel Pepys was an enthusiast, remarking that *Micrographia* was "the most ingenious book that I ever read in my life." Others held a different view-point, distrusting the investigations of Hooke, and other members of the infant Royal Society, and satisfied with, as Galileo had phrased it, the dark labyrinth of their ignorance. When it came to matters of science, Protestant London was no more adventurous than Catholic Rome.

As England's first celebrity scientist, Hooke's demonstrations of experiments at the Royal Society were a sensation. His enormous body of work speaks to boundless enthusiasm and energy, but he was fatigued by endless disputes with lesser inventors and warred with the genius that was the petty and jealous Isaac Newton. Hooke's physical appearance did nothing to boost his public appeal: "He is but of middling stature, something crooked, pale faced, and his face but little below, but his head is large; his eye full and popping, and not quick. A grey eye."[8] More draining than his feuds with fellow scientists was the public humiliation delivered by a play written by Thomas Shadwell titled, *The Virtuoso*.[9]

Shadwell's play, published in 1676, was a typical Restoration comedy, filled with sexual innuendo, cross-dressing, and social satire. The Virtuoso is Robert Hooke, veiled thinly as Sir Nicholas Gimcrack, an ill-fated character who is detested by his nieces, loses his wife, and is threatened by "engineers, glassmakers, and other people" whose livelihoods have been undermined by recent scientific discoveries. Gimcrack's fault as patriarch is to have dedicated himself to experimentation and invested in scientific instruments rather than his family: his niece Clarinda complains, "[he] has spent two thousand pounds in microscopes to find out the nature of eels in vinegar, mites in a cheese, and the blue of plums which he has subtly found out to be living creatures."[10] Hooke discussed eels (nematodes) and mites in *Micrographia*, and Shadwell's "blue of plums" is a reference to his investigations on "blue and white and several kinds of moldy spots" that he recognized as "small and variously figured mushrooms." The subtext of Gimcrack's endeavors is that they threatened the order of the day, which— surviving the Great Plague and Great Fire—relied upon tradition, superstition, and faith. The microscope and other infernal contraptions associated with the Royal Society were changing the way that literate people viewed and made sense of their lives. It was Hooke's misfortune to attend the play. In his diary he wrote, "Vindica me deus" (God grant me revenge).[11]

Most of Hooke's microscopic observations provided readers with surprising views of familiar objects, animate and inanimate, including the flea, linen, snowflakes, and minerals. Unfamiliar exceptions were the fossilized shells of foraminiferans picked out from chalk, and living fungi. The plate showing the spore-producing stalks of a *Mucor* sprouting from a moldy sheepskin book cover, and pustules of the rust, *Phragmidium*, on rose leaves are the earliest published illustrations of living microorganisms. These are, arguably, the most important pictures in *Micrographia* because they show unimagined things, life invisible until 1665.

Hooke's contemporary, Anton van Leeuwenhoek, went much further in his investigations on the microbial world, being the first to describe bacteria, including large, crescent-shaped selenomonad cells from his teeth, a variety of protists, and yeast from beer.[12] Leeuwenhoek used single-lens microscopes of his own design. He fabricated hundreds of these instruments by sandwiching a tiny glass lens between a pair of thin silver or brass plates with a milled aperture. The specimen was held on a spike whose position was controlled by screws. The viewer held the specimen side of the microscope toward the light and squinted from the other side through the brightened lens. The lenses were made by melting thin glass tubes to form "pips" with a diameter of 1 millimeter, whose ellipsoidal shape allowed a wider field of view than a perfect sphere. Lenses in surviving microscopes approach a 300-fold magnification, making the large bacteria from Leeuwenhoek's decaying teeth appear one millimeter, or so, long.[13] With the inadvertent loss and advertent thieving of most of his instruments, and Leeuwenhoek's secrecy about his techniques, it is certainly possible that some of his microscopes were capable of higher magnifications.

We know that Leeuwenhoek employed a variety of methods to hold his specimens in front of the lens. Dried objects could be speared upon, or glued to, the specimen spike; insect muscles were dried onto slips of mica or glass, and pond water containing infusoria and animalcules (bacteria, protists, and microscopic invertebrates) were drawn into capillary tubes whose curved surfaces further magnified the contents. The Dutchman also looked at human tissues, illustrating red blood cells, bled from his maid, and spermatozoa. While I had always assumed that Leeuwenhoek engaged in a selfless program of masturbation for his scientific cause, scholars of onanism suggest that a willing, perhaps enthusiastic, medical student provided the necessary samples.[14]

That blood was full of blobs that resembled yeast cells in beer and semen swam with mini tadpoles, were discoveries that left medical practitioners of seventeenth century medicine dumbfounded. There was an understandable tendency to conflate the cells of our tissues with the independent cells of the bacteria and protists. There was a uniformity to nature deconstructed with the first microscopes, with everything appearing as globules, vessels, and fibers. Other microscopists described minute worms and insects swimming in blood, and the great Italian scientist, Marcello Malpighi, wrote about "red atoms" pulsing through the capillaries of frogs. They imagined that they had discovered the common building blocks of life, which, to a first approximation, they had.

While microscope designs proliferated and optics improved in the eighteenth century, knowledge of the microbial world increased very slowly. George Adams was a "Mathematical, Philosophical, and Optical Instrument-Maker," who "invented, made, and sold" beautiful microscopes from his shop on London's Fleet Street. These included a "New Universal" microscope with multiple lenses fitted into a rotating disc, which was a forerunner of the modern compound microscope with its nosepiece cluster of objectives. Adams detailed his inventions in his *Micrographia Illustrata* of 1746, and illustrated animalcula (synonymous with infusoria) in a variety of fluids, including plant extracts.[15] Many of the protists and invertebrates are recognizable, but a few seem quite hallucinogenic, including the head of a bemused gentleman sprouting six legs, and a microscopic goldfish (Figure 10). Adams copied these, and many other pictures from earlier works, and even stole the beautiful head of the hoverfly from Hooke's *Micrographia*. Evidently, the English scientific renaissance died with Hooke and Newton.

On the continent, significant discoveries with microscopes were made by a handful of investigators who used their instruments as experimental rather than purely observational tools. Two of these pioneers were the Florentine botanist, Pier Antonio Micheli (1679–1737), and Swiss zoologist, Abraham Trembley (1710–1784). Both figures, who are obscured by the greater flowering of biology in the nineteenth century, deserve recognition for their creativity and tremendous determination during an otherwise lifeless interlude in the history of biology.

Micheli's magnum opus, *Nova Plantarum Genera*, published in 1729, is an unappreciated masterpiece of biology.[16] The ostensible purpose of Micheli's book was to showcase new genera of plants.

(a)   (b)

FIGURE 10 Illustrations by French microscopist Louis Joblot. (a) Microscope with lens mounted in a ball-jointed arm. (b) Bemused infusorian gentleman observed in water sample. London instrument manufacturer George Adams reproduced this unusual portrait and wrote, "All the Surface of its Back is covered with a very fine Mask in Form of a *human Face* perfectly well made, as appears in the Figure. It hath three Feet on each Side, and a Tail coming out from under the Mask."

Source: L. Joblot, *Observations d'Histoire Naturelle Faites avec le Microscope* (Paris: Briasson, 1754–1755).

Its plates are every bit as thrilling as the renderings of insects in the *Micrographia*. The first shows the liverwort *Marchantia*. Micheli's use of the microscope is immediately apparent in the details of this plant's air pores, gemma cups, and sporangia that hang from stalked sex organs resembling little umbrellas. Illustrations of hornworts and other simple plants follow, then flowering plants including orchids, followed by studies on 900 fungi. Micheli regarded fungi as primitive plants and described the spore-shedding gills and tubes of mushrooms as flowers that lacked petals. His wondrous portraits of these organisms ranged from the prettiness of cup fungi to the dripping impudence of the phallic stinkhorn. Using his microscope he figured the spores of the larger fruit bodies and those shed by

Mucor, recalling the earlier appearance of this spoilage microbe in *Micrographia* (Figure 11a). He was first to describe *Aspergillus* and *Botrytis*, showing their formation, respectively, of chains and grape-like clusters of spores.

But *Nova Plantarum Genera* was far more than a taxonomic catalog: it included detailed descriptions of experiments. Most significant was Micheli's demolition of spontaneous generation by demonstrating that leaf piles seeded with the microscopic spores of a particular fungus would yield mushrooms of the same organism months later. He did an accelerated version of the same

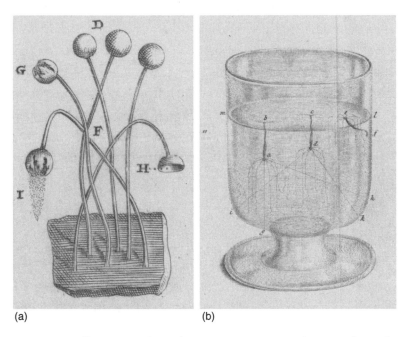

(a)   (b)

FIGURE 11 Illustrations from the most important studies in eighteenth-century microscopy. (a) *Mucor* (Opisthokonta), illustrated by P. Micheli, *Nova Plantarum Genera, Iuxta Tournefortii Methodum Disposita* (Florence: Bernardi Paperinii, 1729). (b) *Hydra* (Opisthokonta), from A. Trembley, *Mémoires, Pour Servier à L'Histoire d'un Genre de Polypes d'eau Douce, à Bras en Forme de Cornes* (Leiden, the Netherlands: Jean and Herman Verbeek, 1744).

experiment by collecting and spreading *Mucor, Aspergillus,* and *Botrytis* spores on pieces of fruit using a soft brush. The microfungi formed new generations of spores within a few days. These experiments were beautiful in their simplicity and anticipated the monumental work of Pasteur by a century.

The German botanist Dillenius said that Micheli was "full of spite all his life."[17] Separating cause from psychiatric effect is difficult after 300 years, but the Italian wasn't treated kindly by his peers. Referring to Micheli's lack of a university degree and lifelong struggles with money the Swiss anatomist Albrecht von Haller libeled him as "illiteratus et pauper." Micheli died in 1737 after contracting pleurisy during a collecting trip in northern Italy. His immediate successors had no interest in microscopy and the consequence of his discoveries was ignored. Despite his neglect by fellow botanists, however, Micheli's life was celebrated in Florence: he was buried in the Church of Santa Croce, in good company with Michelangelo, Galileo, Dante, and Machiavelli; the street outside the Orto Botanico di Firenze was named Via Pier Antonio Micheli, and a beautiful statue of Antonio, holding the *Nova Plantarum Genera* (the title chiseled on its spine), was placed in a niche next to Galileo in the spectacular courtyard outside the Uffizi Museum.

Abraham Trembley enjoyed a better reputation in life.[18] His *Memoirs Concerning the Natural History of the Polyps,* published 1744, introduced species of *Hydra,* tentacled predators related of jellyfish and corals:

> The little creature whose natural history I am about to present has revealed facts to me which are so unusual and so contrary to the ideas generally held on the nature of animals, that to accept them demands the clearest of proofs.[19]

Hydra species are a few millimeters long, and Trembley used a lens mounted on a ball-jointed arm for observing polyps adhering to the inside of the powder jars in which they were farmed (Figure 11b). He noticed that they congregated on the sunlit side of the jars and was surprised by their violent contraction when disturbed. If these organisms were plants, he speculated, the viability of cuttings would settle the question. Recognizing himself vivisector rather than horticulturalist—"I expected to see these cleaved polyps die"—Trembley snipped the animal in half with fine scissors. After a few hours, "I saw it [the anterior end] move its arms"; the next day, "I saw it take a step," and, of the tail end, he exclaimed, "Who would have imagined that it would grow back a head!"

Trembley's dissections caused a sensation. The first reports of whole animals regenerating from severed parts were met with incredulity. To counter skepticism Trembley and his friend René-Antoine Ferchault de Réaumur (inventor of a pre-Celsius thermometer scale) gave live demonstrations of the experiments to excited audiences. Interest spread among fashionable society in France. Madame Geoffrin, "one of the wisest and wittiest women of the eighteenth century," visited Réaumur's study for a firsthand account of the experiments and reported her findings to the philosophers in her Paris salon. Wrestling with the origin of life, Diderot entertained himself with the possibility of "human polypi" on other planets, "Males splitting up into males, females in to females," and "A society of men formed and a whole province populated out of the fragments of a single man."[20] The *Hydra* had established itself in European consciousness. Applause came from London too, and Trembley was elected to the Royal Society the year before his investigations were published. With celebrity came inevitable derision. Recalling Hooke's evisceration in *The Virtuoso*, Trembley was mocked by Oliver Goldsmith, who wrote about "the puny pedant, who finds one undiscovered property in the polype…and whose mind, like his microscope, perceives nature only in detail." Contemplating the green color of the first *Hydra* species discovered by Trembley, Voltaire considered that the polyp was "more like a carrot or an asparagus than it is like an animal."

Like Micheli, Trembley held conventional religious beliefs and regarded his experiments as an exploration of divine creation. But as the wonders of the microscopic world multiplied, questions about the special nature of humanity were unavoidable. The existence of lives unimagined, everywhere we looked, required a reevaluation of ourselves. *Hydra* wasn't in the Bible. In the previous century, René Descartes had envisaged all animals as machines, with humans alone parasitized by an immortal soul. The mechanistic model of life generated interest in robotics. According to one story of unpromising pedigree, Descartes traveled with a female automaton, Francine, who rested in a chest in his cabin during a voyage. To the ship's captain, Francine worked enough like a woman with a soul—meaning not very much—to make him throw her overboard. More convincing automata were created in Trembley's time, including Jacques de Vaucanson's gold-plated mechanical duck that could be fed and would defecate according to mechanisms that mimicked those of the real bird. Few, if any, philosophers argued that the shitting duck possessed a soul, but many found the polyp more problematic. Trembley's cousin, the naturalist and philosopher Charles Bonnet, was troubled by the implications of the regenerative powers of the polyp: "Shall we attribute a soul to it, or none at all?"[21] Sharing his concerns with the Royal Society, Bonnet wondered, "Where then does the Principle of Life reside in such Worms... Are these Worms only mere Machines, or are they like more perfect Animals, a sort of Compound, the Springs of whose Motion are actuated by a kind of Soul?" If the polyp has a soul, asked Réaumur, was it divisible?

Arguments about the indivisibility of the soul escaped most people, of course. Educated Europeans had looked through microscopes and were aware of a few of its surprises, but the instrument remained an object of humor for some time. The anonymous author of a 1732 pamphlet titled *Female Inconstancy* penned a pornographic poem, *The Microscope,* concerning two daughters who borrow their father's microscope to magnify their brother's penis and their own genitalia:

> At last she fix'd her active Sight
> On the sweet Fountain of Delight;
> When loe! It yawn'd so hideous wide,
> That (burst with laughter) *Sally* cry'd,
> To fill that gap, and end thy cares,
> Would ask more... than there are Hairs.[22]

A more progressive portrait of the microscope appeared in the Restoration comedy, *The Basset Table,* written by the dramatist Susanna Centlivre in 1706.[23] The title of the play refers to Lady Reveller's table where friends play a card game called basset. Her cousin, Valeria, is more interested in science than gaming and could not be less interested in a marriage proposal from a sea captain. She transforms her dressing room from boudoir to laboratory and enjoys dissecting fish and other animals under her microscope. To the extent that she prefers her researches to the more common social pastimes of wealthy women, some of the characters deride her like Sir Nicholas Gimcrack's nieces in *The Virtuoso*. But there is more going on in this play. Science is Valeria's buffer against subordination to men, and it is clear that she has no intention of allowing marriage to stifle her investigations.[24] The aspiring scientist is a proto-feminist.

With *Hydra* we had ventured no farther than our own kingdom, and the true breadth of biological diversity was beyond anyone's comprehension in the eighteenth century. Henry Baker, who wrote *Of Microscopes and the Discoveries Made Thereby* in 1742, understood the potential of the microscope for exploring new worlds from one's desktop.[25] He was fascinated by the colonial alga, *Volvox*, a rolling green ball built from as many as 50,000 cells, each equipped with a pair of flagella. Baker's writes about "The Globe Animal" with an evident sense of excitement, something absent from papers in today's scientific journals:

> In the Month of *July* 1745, three Phials full of Water were sent to me from *Yarmouth*, by Mr. *Joseph Greenleafe*, having in them several Kinds of Animalcules unknown to me before.… They all died with me in two or three Days, but in that Time I had Opportunities enough to examine them…Its Form seems exactly globular, having no Appearance of either Head, Tail, or Fins. It moves in all Directions, forward or backwards, up or down, either rolling over like a Bowl, spinning horizontally like a Top, or gliding along smoothly without turning itself at all.… The Surface of the whole Body appeared…as if beset thinly round with short moveable Hairs or Bristles; and 'tis not improbable all their Motions may be produced by some such Instruments, performing the Office of Fins.

Anyone who has looked at a live *Volvox* colony will identify with Baker's description. He described rotifers too, figuring "The Wheel Animal in its several Postures," and "Funnel-Animals"

(*Stentor*), "Bell-Flower or plumed Polype Animals" (*Vorticella*), and "Oat-Animals" (diatoms) whose "Shell is so exquisitely thin." *Amoeba* was another protist that spilled itself into view at this time, severing itself—and its soul?—into a pair of lively daughter cells without any of the recovery time required for the *Hydra*.[26] Baker wrote that these discoveries "serve to shew our Ignorance concerning the real Essence and Properties of what we term *Life*."

In the decades after the discoveries of Hooke and Leeuwenhoek, the microscope made more of the invisible visible and mankind magnified looked no different than the things that wriggled in ponds. Reflecting on the human machine with this newborn clarity we found ourselves made from infusoria and animalcules. For the first time, investigators were beginning to make sense of nature without the encumbrance of religious superstition, but, as I noted earlier, the enterprise faltered in the eighteenth century. Micheli and Trembley were quickly forgotten and the promise of a profound reimagining of ourselves and our place within the rest of nature came to nothing. The cost of this monumental lapse in concentration was another century of confusion about spontaneous generation, until Pasteur sorted things out, followed by another century of continuing insistence upon the privileged rank in creation held by humans.

It is difficult to fathom the cause of this lengthy stasis, but I think that the answer lies in the nature of the investigations pursued after Micheli and Trembley. Part of the reason that Micheli's work was ignored was that the botanists who embarked upon the methodical cataloguing of plants derived a simpler method of classification that made Micheli's work redundant. The formal names that Micheli used for his plants and fungi were snapshot descriptions of the organism.[27] The aforementioned phallic mushroom, the common stinkhorn, offers a fine illustration. Micheli named this, *Phallus vulgaris, totus albus, volva rotunda, pileolo cellulato, ac summa parte umbilico pervio, ornato*, meaning, common phallus (stinkhorn), white bodied, with a rounded shell, and an ornate chambered cap adorned with a navel at its tip.

This ungainly method of phrase naming (*polynomial*) was shattered by the publication of the first edition of the *Species Plantarum* in 1753 in which Linnaeus gave each plant a *binomial* name. The phallic mushroom became *Phallus impudicus*, shortening its formal announcement from eight seconds to one second, and has remained so ever since. Linnaeus had little respect for Micheli's work, and the Swede's dominance of eighteenth-century botany ensured that the Florentine genius was forgotten. This wouldn't have been a significant loss if Micheli had limited himself to taxonomy, but the love of the binomial classification eclipsed everything in the *Nova Plantarum Genera*, and the crucial experiments that refuted spontaneous generation were ignored for 200 years. The obsession with cataloguing inhibited clever investigation, and natural history entered the stamp-collecting phase of its arrested development. As I mentioned in the first chapter, too many contemporary biologists remain there today.

Investigators benefitted from a remarkable improvement in the clarity of the magnified image in the 1820s, when the first microscopes that corrected for spherical and chromatic aberration were introduced.[28] Spherical aberration blurs the magnified image by focusing light passing through the edge of a curved lens onto a different spot from the light transmitted through the middle of the lens. This was addressed in the new microscopes by combining lenses to flatten variations in focal length and by narrowing the beam of light to the center of the lens with a diaphragm.

Chromatic aberration results from the dispersion of light of different wavelengths as it passes through an uncorrected lens, radiating colored halos around magnified objects. This was solved by the fusion of two lenses made from different kinds of glass ("flint glass" cemented to "crown glass") into a "doublet" that causes the rainbow of colors to converge as they pass through the lens. The Abbe condenser, developed for the German manufacturer Carl Zeiss, was a later improvement. This accessory was, and still is, mounted beneath the stage of a conventional microscope and concentrates a sharp cone of light upon the center of the field of view.

These optical advances, coupled with a refreshing commitment to experimentation, allowed Victorian scientists to overcome many of the obstacles to understanding the structure and function of living things. Remaining supporters of spontaneous generation were silenced by Pasteur's demonstration, in the 1860s, of the incorruptibility of sterilized broth as long as it remained isolated from airborne microbes. These experiments also helped support the Germ Theory of Disease, or Pathogenic Theory of Medicine. Incidentally, another French investigator, Bénédict Prévost, had offered experimental proof that a particular microorganism caused a specific disease 50 years earlier.[29] Because Prévost was working on bunt of wheat, caused by a smut fungus, rather than a more newsworthy combination like a little boy dying of rabies (one of Pasteur's triumphs), nobody has heard of him.

Other highlights of nineteenth century microbiology included a brilliant exposition on fungi and bacteria, *Comparative Morphology of the Fungi Mycetozoa and Bacteria*, written by Anton de Bary, and Ernst Haeckel's tree of life showing single-celled "protista" occupying the branches between the plants and animals.[30] Both works were published in 1866 and illustrated the progress in exploring the diversity of microscopic organisms. Commenting upon bacteria in his book, De Bary said, "Nuclei have not as yet been identified in bacteria." The fundamental distinction between nucleated eukaryotes and nucleus-lacking prokaryotes wasn't recognized until the 1930s by Édouard Chatton.[31]

Advanced optics made the work on the single cells of bacteria, protists, and fungi more straightforward, less open to guesswork, making it possible to probe the subcellular architecture of microbes that had been viewed as little more than mobile blobs with earlier microscopes. Experts on all manner of organisms used the new instruments to document unimagined details of cell biology. Among the mycologists who looked at my favorite organisms, I am awed by the stunning illustrations of microscopic fungi published by the Tulasne brothers and the German botanist Oscar Brefeld.[32] The quality of their drawings transcends later photographic images because they were able to combine information from several focal planes, buy focusing up and down, and to incorporate features from numerous individual cells into a single picture.

Experts on other kinds of microscopic organisms were similarly successful in documenting the stunning forms of photosynthetic and non-photosynthetic protists. Most of the bacteria were too small to allow investigators to sort them into anything other than *almost* meaningless groups based on whether the cells were spheres (cocci), rods, filaments, or spirals, and whether they clumped together or swam around singly. Even the highest magnifications, using oil- or water-immersion lenses to enlarge the cells more than 1,000 times, revealed little detail. Distinctions in cell structure were discovered, however, using dyes to reveal differences in the chemical composition and architecture of the cell wall. The classic Gram stain, for example, colors the thick peptidoglycan wall of *Bacillus* and *Clostridium* species, and other Gram-positive bacteria, with crystal violet; the dye is washed from the walls of *Escherichia coli*, and related Gram-negative bacteria, which are then counterstained pink with a different dye. This technique was published in 1884 and remains a useful tool today.

At the end of the nineteenth century, natural historians had shed the 2,000-year-old whimsy that all biology was animal and plant, but continued to sort everything into the realms of zoology and botany.[33] Protists that lacked chloroplasts were accorded the status of simple animals and studied by zoologists; photosynthetic protists, or algae, were part of botany, along with bacteria and fungi. Microbiology was beginning to carve a niche for itself and this disciplinary harvest would include most of the bacteria, plus the disease-causing fungi. Botanists would continue to handle the photosynthetic cyanobacteria and call them the blue-green algae. The prokaryotes were divided from Haeckel's protists in the 1930s, with the creation of a fourth kingdom called the Monera, as more and more research showed that there was a fundamental difference in cell structure between the bacteria and everything else.[34] Bacteria were the elementary cells; the cells of protists, plants, and animals were more complex, composite things with nuclei.

The next breakthrough came when the fungi were removed from the plants and given their own kingdom in 1969, a breakthrough of comparable scientific importance to the moon landing in July of the same year.[35] This makeover was proposed by Robert Whittaker, a plant ecologist, who argued that their saprotrophic behavior distinguished the fungi from the solar-powered lifestyle of the plants. The logic of this distinction notwithstanding, the fungi continued to be viewed as a botanical oddity and mycologists continued to work in botany departments. Nobody else wanted them. The persistent status of the fungi and their keepers as an oppressed minority is evidenced in the brief chapter on "mushrooms and their allies" included in every introductory textbook of botany.

Whittaker's five-kingdom arrangement of life persisted until research by Carl Woese argued for separation of the prokaryotes into the bacteria and the archaea.[36] Rather than looking at the structure of prokaryotes, Woese concerned himself with the DNA sequence of a single gene that encodes part of the structure of the ribosome. Ribosomes are organelles made from complexes of RNA and protein that function in the synthesis of new proteins. All cells contain related sets of genes that code for the RNA portions of ribosomes. Changes in the sequences of ribosomal RNA genes, like all other genes, arise through evolutionary modifications as millions of years slip by. This means that the farther apart one organism is from another—you compared with a hamster, for example, versus you and *Amoeba*—the greater the divergence between the sequences of their versions of these genes. Sequencing these genes in prokaryotes, Woese found that the differences in code between archaea and bacteria were so great that they must have diverged from one another a very long time ago and pursued separate evolutionary paths. Following this deduction, Woese argued that archaea, bacteria, and eukaryotes represented three "aboriginal lines of descent."

Maintaining the earlier subdivision of the eukaryotes into four groups, the invention of the archaea bestowed six groups upon biology. The resulting six-kingdom organization is still taught in college biology classes despite having little to recommend it beyond its service in multiple-choice tests. The more important consequence of Woese's genetic inquiry was to expose the true variety of microbes, and to relegate plants and animals to little twigs on an evolutionary tree that kept expanding. Critics fought the cleavage of the prokaryotes.[37] Their blunder was to assume that evolution split groups along clearly visible lines: after all, they reasoned, one prokaryote looked much the same as another and mere differences between genes could not inform a reliable classification. The eukaryotes were misread with equivalent myopia, and biology is still playing catch-up with the fact that evolutionary innovations among the protists swamp every measure of plant and animal diversity.

Before we leave the history of biological thought for a dip in the sea, I want to return to Galileo. The astronomer received John Milton at his house in Tuscany in 1638. Milton was 30; Galileo was elderly and blind. Twenty years later, now sightless himself, Milton began to compose *Paradise Lost*. Early in the first book of the great poem he recalled his grand tour of Italy and likened Satan's shield to the moon seen through Galileo's telescope.[38] The quotes from the poem that grace each chapter of my book show Milton's fascination with the workings of nature. The extract for this chapter, from Book III, "that I may see and tell of things invisible to mortal sight," was not, unfortunately, a reference to microscopy, but reflects the poet's plan to address "the ultimate question of life, the universe, and everything" (to quote from a lesser bard called Douglas Adams). In many ways, Milton was handicapped by an Aristotelian worldview, but he was an ardent protector of free speech and of investigation unencumbered by religious censorship. We are blessed to be living in the twenty-first century, when science has clarified our picture of reality beyond Milton's, or Galileo's, wildest dreams. Fortunately, or unfortunately, depending upon one's perspective, our view of ourselves is going to change in the next decades as much as it did during the Restoration. This shift in thinking requires us to embrace a fully objective exploration of nature, one in which we almost vanish from view.

3

Leviathan

Hugest of living Creatures, on the Deep
Stretcht like a Promontorie sleeps or swimmes,
And seems a moving Land, and at his Gilles
Draws in, and at his Trunck spouts out a Sea.
        —Milton, *Paradise Lost, Book VII*

The largest eye belongs to the biggest mollusk, *Mesonychoteuthis hamiltoni*, the colossal squid of Antarctic waters. The paired orbs of this cephalopod are one foot in diameter and hold lenses as big as oranges. The retina is oriented toward the incoming light which is focused by moving the spherical lens forwards and backwards (our retinas point toward the brain and the shape of the lens changes to focus light upon the rods and cones). The extraordinary sense organ of this squid is adapted for vision under the low light conditions of its deep feeding grounds where it must detect the flick of a fish tail or the menacing shadow of a diving sperm whale.

The smallest eyes are also found in the sea. They develop inside the single cells of the warnowiid dinoflagellates, *Erythropsidinium*, *Nematodinium*, and close relatives, bulging from the cell surface, presenting a crystalline lens that focuses light onto a wine-dark photoreactive cup (Figure 12).[1] The dinoflagellate peepers resemble animal eyes, but we don't know what functions they perform. They may give these odd protists the edge as predators, detecting their quarry rather than waiting to bump into it. One idea is that the eye operates as a range finder, improving the accuracy of prey capture using barbed nematocysts or other cytoplasmic weapons available to the warnowiid.[2] Or perhaps they alert the cell to a dangerous shape, the outline of enemy crustaceans that have pestered dinoflagellates for tens of millions of years. Being an eye housed in a single cell, image processing is going to be limited to no more than the flip of a binary switch: the algorithm reading something like, *when 75 percent of the retina is darkened, swim away from source of shadow ASAP*. The eye has a 30 degree field of view, which would be

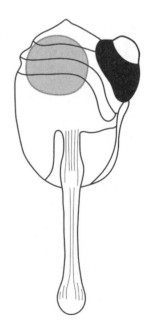

FIGURE 12 Warnowiid dinoflagellate *Erythropsidinium* (Alveolata) with ocelloid formed by dark photoreactive cup beneath a hemispherical cystalline lens. The pendulous tentacle is thought to be used for feeding.

58

filled by a millimeter-long crustacean at a distance of 2 milli-meters, giving the protist a few thousandths of a second to es-cape the predator's "feeding vortices"—water currents that suck plankton toward their mouthparts.[3] Alternatively, the eye may have nothing to do with vision. A miniature "burning glass" seems plausible, the dinoflagellate focusing the sun's rays upon the pigment molecules in its retina, heating the cup to catalyze biochemical reactions vital to its unseen behavior. Or, perhaps it works like a chloroplast, absorbing photons and manufacturing chemical energy.[4]

Whether the warnowiid eye is a hunting, warning, warming, or fueling device it is as wondrous as the massive orbs of the squid. Very few scientists have seen a *colossal* squid, also known as the giant cranch, but the existence of this Leviathan, or the related *giant* squid, is familiar even to those indifferent to the charms of natural history.[5] None of the more than 1,500 species of dinofla-gellates demand this kind of star power. Their greatest exposure comes when bioluminescent species brighten the wake behind a cruise ship, or form a shimmering blue halo around a tourist swimming within a coastal bloom. Despite their low billing, dino-flagellates are at least as interesting as cephalopods, and, as I'm going to explain, of much greater significance for planetary health.

Dinoflagellates do a lot of the photosynthesis in coastal waters, and are the largest of the single-celled organisms caught with a plankton net. Thirty years ago I learned about these prot-ists during Frank Round's lectures on algal biology at the Uni-versity of Bristol.[6] Frank taught an annual field course on marine algae at the Plymouth Marine Laboratory in England which al-lowed us to study fresh samples of plankton brought to dock by the lab's research trawler. The catch of the day was also anticipated by the great zoologist, J. Z. Young, a visitor to the Plymouth labs, who had discovered the giant axons of squid used in the study of nerve impulses in the 1930s. He wore a blood-stained lab coat and dangled a long metal rod from a belt fashioned from string. Frank explained that this was a "pithing rod" that was pushed into the spinal column of fish, and other unfortunates, slid forward into their brains, and used to dispatch the writhing creatures in preparation for dissection. While Professor Young rendered his vertebrates, we marveled at the cells of *Ceratium* and *Peridinium*, beautiful armored dinoflagellates that swam spirals in the spot-lit drops of salt water on our microscope slides (Figure 13).

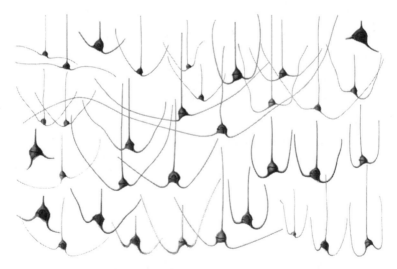

FIGURE 13 Collage of the marine dinoflagellate *Ceratium tripos* (Alveolata). *Source*: G. Karsten, *Das Indische Phytoplankton* (Jena: Gustav Fischer, 1907).

Dinoflagellates are the fastest algae, whizzing up and down the water column at 1 meter per hour, positioning themselves at the perfect height to capture light according to the brightness of the day, and pinpointing concentrations of dissolved vitamins and other nutrients. As I explained in the first chapter, dinoflagellates have absorbed a variety of photosynthetic protists during their evolutionary history to create chloroplasts; species that lack chloroplasts have dispensed with their partners, or never captured them in the first place, and operate as predatory members of the plankton.[7] We can assign the photosynthetic ones to the *phytoplankton* and the predatory ones to the *zooplankton*, as long as we recognize that these are categories of function rather than meaningful taxonomic arrangements. After all, some dinoflagellates are photosynthetic *and* predatory. These are called *mixotrophs*. Irrespective of food source, these free-living flagellates are *pelagic* microbes, organisms that live in the upper part of the water column.

Pelagic dinoflagellates are ignored in introductory biology classes, but their symbiotic relatives in corals get a mention as the photosynthetic livelihood of reefs. The coral dinoflagellates, or *zooxanthellae*, belong to the genus *Symbiodinium*, which fuel their animal associates with glycerol.[8] *Symbiodinium* clones can occupy corals over wide areas, but there is greater genetic variation between reefs. Some versions of the symbiosis appear to be parasitic, with the flagellates donating 90 percent of their photosynthetic sugars to the animal and getting little in return.[9]

The genetics of *Symbiodinium* are very complicated and the genus has been split into a number of distinctive lineages, or *clades*, rather than species for now. Rising sea temperatures are linked to an increasing frequency of coral bleaching events in which the dinoflagellates leave their polyps and cause their starvation. Some clades seem to offer greater resistance to bleaching than others, which is one of the reasons that researchers are so interested in the genetics of these algae. The photogenic brilliance of unbleached coral reefs and the appalling specter of their destruction demand our attention, but coral reefs are, like the mammals on the Serengeti, distractions from the larger microbial domination of life. Coral reefs cover a little less than 300,000 square kilometers and there are 325 million square kilometers of open ocean, and we'll spend most of this chapter on the pelagic slice of life that lives out there.[10] Microbes rule the bottom-dwelling, or *benthic*, ecosystems too, including the cold sea bed and hot ocean vents; we'll visit the abyss later.

To approach a meaningful picture of marine biology, we need to put aside the things studied by zoologists. A sushi bar to end all sushi bars will foster the necessary thought experiment. Every morsel of marine muscle must be eaten in this last supper: all the hagfish, lampreys, sharks, rays, and bony fish are diced, rolled in sticky rice, wrapped in seaweed, kissed with soy sauce, and swallowed; the red meat from whales, dolphins, manatees, and walruses works well as sashimi and sea turtles make soup; all the oysters slip down with the assistance of cold white wine, all the squid are crunched calamaried; orange sea urchin gonads make a sloppy topping for sushi rolls and jellyfish can be fried. Crabs and lobsters are dispatched after boiling, along with the related sea spiders, barnacles, and fish lice. This is a lot of food: fish, great whales, and Antarctic krill alone weigh more than 1,000 million tons. That leaves the sponges and comb jellies, penis worms and other worms, and exotics like mud dragons, but most of the gustatory labor is over and the ocean is much clearer for it.

Now we can turn our full attention to the 90 percent of living things in the sea that cannot be seen without a microscope.

The ubiquitous ocean dweller, the cell that is probably more numerous than any other, is a cyanobacterium called *Prochlorococcus*.[11] There are one octillion or more of them in the sea—as many blue-green cells floating in the sunlit waters of earth as there are atoms organized in a human body.[12] In the open ocean where levels of dissolved nutrients are relatively low, *Prochlorococcus* is responsible for a lot of the photosynthetic activity. This means that it is a big player in global nutrient cycles, incorporating carbon from the growing quantity of carbon dioxide in our atmosphere into the substance of its cells. So much carbon is fixed by this microbe, and so much oxygen is emitted, that if *Prochlorococcus* quit working the planet would warm intolerably and the air breathing portion of life would suffocate. It is possible that other microbes would fill the breach and pull down the carbon dioxide and exhale enough oxygen to keep things running in our favor, but this would take a while. If *Prochlorococcus* failed, I wouldn't bank on a comfortable retirement, and as for the grandchildren's quality of life...not much. We are yoked invisibly to this bacterium.

Prochlorococcus and all the other photosynthetic bacteria and protists are responsible for half of the biological uptake of carbon dioxide on earth, equivalent to all the photosynthesis by bacteria, algal protists, and plants on the land. Other cyanobacteria bob alongside *Prochlorococcus*, fertilizing the water by fixing atmospheric nitrogen into ammonium. *Trichodesmium* is one of the important ones, whose widespread blooms are called "sea sawdust."[13] Non-photosynthetic plankton feed upon *Prochlorococcus* and other microbes, consuming oxygen and releasing carbon dioxide, and are incapable of reducing the amount of carbon dioxide in the atmosphere.

Photosynthesis consumes carbon dioxide and produces oxygen; respiration consumes oxygen and produces carbon dioxide. The balance between consumption and production is in constant flux. Photosynthetic activity tends to be greatest close to the ocean surface and oxygen production can exceed consumption here. The situation is reversed at greater depths, where respiration outstrips the modest release of oxygen from shade tolerant algae that stir in the gloom. Researchers studying the chemistry of the open ocean often find that consumption exceeds production.[14] The ocean appears doomed by deficit spending. This obviously unsustainable situation seems to be resolved by the phenomenon of periodic blooms, or population explosions, of additional photosynthetic cells that generate huge bursts of oxygen that sustain the next cycle of seemingly relentless feeding and planktonic decay.

Marine diatoms are supremely effective at removing carbon dioxide from the atmosphere and generating oxygen from seawater. They flourish in water that is richer in nutrients than the leaner locations colonized by *Prochlorococcus* and turn the surf brown when they bloom. By absorbing 20 billion tons of carbon per year—about one fifth of the global total—they are a greater refrigerant on this warming planet than tropical rainforests.[15] More carbon is held within the tissues of the forest trees, but much of this is sequestered in woody tissues that are metabolically dead. By contrast, there is no more to a diatom than its microscopic cell, and it is drawing down carbon dioxide every sunbathed minute of its life.

There are no cells more beautiful than diatoms in the universe. Each is a lidded glass dish with girdle bands intervening between dish and lid. Shapes vary from canoe to rounded-corner triangle to oval to perfect disc. Our impression of diatom beauty derives from an innate response to the symmetry of the cells and their elaborate piercings. (Diatoms stimulate the same circuitry, I assume, as sea-shells and flowers: we always search for and feel rewarded by the discovery of patterns. This may explain why one can spend so long contemplating abstract art, seeking and failing to find, for example, pattern in a Jackson Pollock.) This regularity of form appealed to Victorian microscopists who arranged diatoms on glass slides using boar bristles and cat whiskers, fixed the arrangements with egg white or Canada balsam, and created a market for these extraordinary creations. Some of this painstaking work served the scientific purpose of documenting diatom diversity from particular locations, but professional mounters found a wider audience for whimsical collages of dancers and floral bouquets assembled from hundreds of cells.

The cell wall of the diatom is a glass shell, also called a *frustule*, composed of partly hydrated silica. Silicon is the second most abundant element in the planet's crust (oxygen is number one), and diatoms proliferate in water with high concentrations of dissolved orthosilicic acid. It seems like an unusual casing for a cell, but this impression may derive from a time when diatoms were the special purview of botanists and the botanical view of the cell wall was prejudiced by familiarity with plant cells walls made from cellulose. As soon as we look beyond plants, we encounter cells walled with all kinds of compounds including non-cellulose polysaccharides, glycoproteins, chitin, peptidoglycans, and chalk. A glass wall seems to be a good option for diatoms because its construction consumes less than one-tenth the energy of a sugar-based wrap.[16]

Each time the diatom cell divides, it fashions a pair of new shell halves, or *valves*, within its cytoplasm, so that each of the resulting daughter cells carries one valve from its mother and makes a fresh one that is smaller than the original. The daughter that inherits her mother's lid is the same size as her parent; the daughter that receives the base is somewhat smaller. (Feminine pronouns are conventional when describing dividing cells, rooted in an era, perhaps, when most naturalists were clergymen.) This means that the average cell size in the growing population shrinks with each division. Because frustule synthesis is so efficient, diatoms can divide at a feverish rate when conditions are perfect, and this results in the depletion of dissolved silica as the cells get smaller and smaller. This lemming-like performance is dodged by the marvel of diatom sex.

Planktonic diatoms with a circular or smoothed triangular form, called centric diatoms, produce male cells that release sperm and female cells that cradle an egg. Life at sea in a floating greenhouse presents a novel challenge to intimacy that is solved by separation of the valves of the male cell, which allows a drop of semen to spill into the water, and the female accommodates by loosening her girdle. Propelled by a single flagellum, a sperm cell dissolves into the midriff bulge bared by shifted girdle bands, fertilizes the egg within, and forms the next full-size cell that sires a clonal lineage of ever-shrinking kin. Sex restores cell size because both halves of a new shell are formed with maximum dimensions around the fertilized egg.[17]

The discarded shells of the parents, along with the frustules of dead diatoms, drop through water column, glint for the last time in the weakening sunlight, and sink into the depths. Theirs is a slow descent, the viscous drag of water on the tiny shells opposing the gravitational imperative. The rate of descent is no faster than one meter per day for single shells, but aggregation of organic particles adds to a heavier marine snow that pours toward the abyss. Most silica shells dissolve as they drop though the water; only two or three percent complete the journey. Yet diatom deposits with thicknesses of hundreds of meters form during the passage of millions of years. An 80-meter thick deposit from the late Miocene in Peru contains exquisitely preserved fossils of hundreds of whales.[18] The current model for these gorgeous burials—even the microscopic detail of the baleen is preserved from some animals—is that the carcasses sank into the deep powder of uncompressed diatoms that resettled as a scavenger-proof shroud.

Deposits of diatomaceous earth, or diatomite, are exposed in many parts of the world. The United States is the biggest producer, with an annual excavation of 600 million tons.[19] The mineral is used in filters, as an additive in paints and cosmetics, and as a mild abrasive. Many of the commercial sources of diatomite were deposited in freshwater lakes, but others, including the huge Lompoc Plant in California, were formed by marine diatoms that bloomed in the Miocene. The presence of deep deposits of pure diatom shell in these mines may help us comprehend the scale of the life and death of planktonic diatoms in today's oceans. Geological manifestations of immense marine productivity are more obvious than estimates of carbon burial or cell numbers. Another group of protists offers an even clearer illustration of the microbial authority over life at sea: the coccolithophorids.

The White Cliffs of Dover are one of Europe's natural wonders, their sight conjuring for me an immediate feeling of tremendous

Britishness, pride in the Royal Air Force (The Few), and an over-powering desire for a cup of the brightest tea in a flowery bone china cup. The 100-meter cliffs expose layers of the soft chalk that was deposited in the Late Cretaceous, 80–85 million years ago. The planet was in one of its extreme greenhouse phases then: carbon dioxide levels were four times higher than today (six times above pre-industrial concentrations), 30 percent of the atmosphere was oxygen (versus 21 percent today), and sea surface temperatures were around 100 °F, offering a comfortable bath for the mosasaurs and other sea monsters. Sea levels were 200 meters above today's mark. The accumulation of chalk was formed by the sedimentation of the calcareous scales that envelop the cells of the coccolithophorids. Creation "geologists" are horrified by the White Cliffs of Dover: their websites discuss the changing vicissitudes of life during the Noachian Flood, but those unscalable cliffs must look like an awfully compelling argument for an ancient earth to all but the most delusional site visitors.[20]

Like the diatoms, coccolithophorids are photosynthetic algae, but they are members of the hacrobian supergroup, a giant phylogenetic leap away from the stramenopile diatoms. Both algal types produce an elaborate cover, but their chemistry and construction bear no similarity. The scales on the coccolithophorids, called *coccoliths*, are made from crystals of calcium carbonate. They look like miniature shields. By surrounding themselves with 15 or more of the coccoliths, the algae resemble the shield wall defense of Roman Legions called the *testudo formation* (Figure 14). Both algae and infantry look a bit scruffy, but armor is more important than aesthetics in biology and war. While protection seems like a logical function for the scales there may be other advantages to their elaboration. One suggestion is that each scale acts as a lens, focusing light upon the chloroplasts as the cells drift with the water. Each scale is assembled individually within the cell and extruded onto its surface.

In a microscopic study of a widely distributed species of coccolithophorid, a group of investigators found that assembly of a single scale takes about 3 hours.[21] They also documented a 2–3 minute secretion process in which the algae "exhibit continuous visceral contractions and slow rotary movements" as they extrude the plate onto the cell surface. A time-lapse video of this spectacle looks like a feat of monumental constipation relief.

FIGURE 14 The coccolithophorid *Emiliania huxleyi* (Hacrobia).
Source: Debby Mason (www.debbymason.com), with permission.

Whatever the function for the algae, the formation of cocco-liths is a vital process in climate control that is, in turn, affected by shifts in climate. When carbon dioxide dissolves in sea water it forms carbonic acid (H_2CO_3); each molecule of carbonic acid splits, spontaneously, to produce a bicarbonate ion (HCO_3^-) and a proton, and a further reaction splits the bicarbonate ion into a carbonate ion and another proton. These are reversible ioniza-tion reactions and their rates are controlled by multiple factors, including the temperature and alkalinity of the seawater. Carbon dioxide is dissolved at relatively low concentrations in the water and bicarbonate ions dominate the chemical mix. Coccolitho-phorid cells absorb bicarbonate ions and these combine with cal-cium to produce the scales according to the following reaction:

$$2HCO_3^- + Ca^{2+} \rightarrow CaCO_3 + CO_2 + H_2O$$

This reaction shows the fate of a pair of bicarbonate ions that ori-ginated as a pair of carbon dioxide molecules that dissolved in the sea. For each pair of bicarbonate ions, the protist releases one molecule of carbon dioxide and stuffs another carbon atom into its scales. Under conditions when the algae are very active, during the expansive phases of a bloom, photosynthetic activity mops up huge quantities of carbon dioxide. Coccolithophorid blooms, or population explosions, cover vast areas of the ocean and are easily visible as underwater clouds in satellite images.[22] Dense blooms cover more than one million square kilometers of ocean every year. The combination of photosynthesis and scale forma-tion means that these blooms operate as enormous carbon di-oxide sponges. When the activity of the bloom slows, however, the algae tend to shift toward net production of the evil gas. The only process that takes carbon out of the loop forever is the sedimentation of scale-laden coccolithophorids all the way to the ocean floor. And the algal blooms have other important effects upon climate. The white clouds of chalk-covered cells increase the reflectance of light from the ocean, and when the cells die they release dimethylsulfide, a volatile sulfur-containing compound that aids cloud formation by acting as a seeding agent. Reflectance and cloud formation are processes that promote global cooling.

In light of the significance of the coccolithophorids in controlling the planetary climate, it is not surprising that investigators have been concerned by the effects of ocean acidification upon algal vitality. Before we began burning fossil fuels, the pH of the oceans was 8.2; in the last 150 years or so, acidification has dropped pH to 8.1. The pH scale is logarithmic, so this translates to a 20 percent decrease in pH. This acidification is linked to the increase in carbon dioxide levels in the air through the reactions described earlier: protons are released from the reactions that occur when carbon dioxide dissolves in water. As seawater become more acidic, the proportions of the different forms of dissolved carbon change, and this has the potential to disrupt the process of scale formation.

Some studies have shown that algae form weakened, crappy-looking scales when pH drops; others have shown that coccolithophorids can form even denser scales under the same conditions. The current working model accepted by many experts is that different coccolithophorid species respond differently to acidification, and that the longer-term effect of such a major change in ocean chemistry will be to alter the relative abundance of different types of these algae.[23] This conclusion is supported by analysis of the fossil record. Remember that the chalk cliffs in

71

England were formed during a period of extreme greenhouse conditions—this shows that coccolith formation can adapt to far higher levels of carbon dioxide than we are likely to achieve in the near future.

It would be a mistake, however, to embrace a Panglossian stance to climate change and assume that the algae will adapt and that everything will be fine in this best of all possible worlds. Human activity is altering atmospheric chemistry at an alarming clip, and adaptations that occurred millions of years ago to allow chalk formation may have taken a very long time to play out. Short-term changes in environment can wipe out a lot of species, and the sensitivity of marine microbes to subtle changes in atmospheric chemistry should make everyone wary.

Before the 1970s, textbooks showed energy flow from diatoms that made the food, to copepods and krill that filtered them from the water, to fish and whales that completed the chain.[24] The emphasis was zoological. John Steinbeck described a collecting trip with biologist Ed Ricketts in the 1940s in his delightful memoir *The Log from the Sea of Cortez*.[25] During the six-week trip, Ricketts and his assistants collected and pickled 500 animal species from the Gulf of California including crabs, sea anemones, limpets, and barnacles. There was little interest among oceanographers at that time in the microscopic residents of the sea. Specialists who studied algae, called *phycologists*, were engaged in detailed descriptions of the structure of photosynthetic protists, and few scientists troubled themselves with the ecology of these single-celled organisms. Study methods hadn't changed since the nineteenth century; a resurrected Charles Darwin could have worked alongside Ricketts and Steinbeck without any additional training. The focus had always been on the animals that I turned into sushi.

This picture has been inverted in the last quarter century, with the recognition that microbes account for most of the biomass in the oceans and run the entire operation. We know now that microbes control nutrient cycles in the oceans, operate complex food webs that support the traditionally charismatic organisms with flippers and flukes and retirement plans, and ameliorate the damaging effects of humanity. Microbial oceanography has become a very important scientific endeavor and its study methods would astonish Ed Ricketts and his crew.[26]

Modern techniques include remote sampling of plankton using satellites, automated plankton identification using submerged cameras, and the truly revelatory methods of molecular biology. In a 2004 study, genomic researchers used a method called *shotgun sequencing* to study microbial populations in the Sargasso Sea near Bermuda.[27] Rather than sequencing genes from particular organisms sorted from the water, this metagenomic method provides sequence information derived from all of the DNA within a sample. The researchers began by filtering 200 liters of water from four different sampling locations to concentrate bacterium-sized particles. The next step was to isolate the DNA, chop it into pieces, and insert these pieces into lab bacteria (the molecular biologist's workhorse, *Escherichia coli*) to create a living repository for the information extracted from the seawater. The accessions in this *DNA library* were then sequenced, ordered via computer into more meaningful lengths by combining overlapping sequences, and compared with existing databases of genes to determine their likely function.

The study resulted in an unprecedented haul of 1.2 million new genes that showed little or no similarity to known genes. Part of the reason for this bounty was that the shotgun sequencing method removes all reliance upon culturing organisms: to be in-cluded in the inventory, a bacterium simply had to have been living in the water. Complementary analysis of the ribosomal RNA genes from the samples provided information on the organisms filtered from the water. Proteobacteria were most common, but photosynthetic cyanobacteria, Gram-positive firmicutes, seven additional groups of bacteria, plus archaea, were represented in the Sargasso Sea. Specifying a particular threshold of genetic difference as a somewhat arbitrary measure of species richness, the investigators estimated that they had amplified DNA from at least 1,800 species of marine bacteria and archaea.

Protists, with their much larger genomes, present a greater challenge for molecular fisherwomen and fishermen. The genome of the cyanobacterium *Prochlorococcus* encodes 2,000 genes, which is eclipsed by the 11,000 genes in the nucleus of the diatom *Thalassiosira* and estimated 39,000 genes in the coccolithophorid *Emiliania*.[28] One technique for exploring eukaryote diversity is to amplify specific sequences from seawater samples that are diagnostic of the supergroups. This approach nets sequences that fall outside all of the recognized groups, including "picobiliphytes," plus novel stramenopiles and unidentified marine fungi. Even among the best-known groups of marine protists, very few sequences can be reconciled with organisms that have ever been cultured in the lab. Up to 98 percent of the sequences retrieved from Indian Ocean waters in a 2008 study were new to science.[29]

The picobiliphytes have not been cultured, so we know almost nothing about them beyond their genetic signatures. They appear to be distant relatives of the cryptomonad algae and may nestle within the hacrobian supergroup of eukaryotes that also includes the coccolithophorids. To begin to learn more about these organ-

isms, investigators sequenced the genomes of three *individual* pic-
obiliphyte cells collected from seawater samples from the Gulf of
Maine.[30] These cells were sorted from the millions of other planktonic protists using a flow cytometer, their DNA was extracted
and amplified hugely, then chopped into pieces, assembled into a
DNA library, and sequenced. Most investigators had assumed
that these cryptic constituents of the plankton were photosynthetic, but their DNA lacked all traces of chloroplast genes, suggesting that the organisms in this study were heterotrophs. Other
surprises included the discovery that the genome of one of the
cells was infested with genes from a previously unknown DNA
virus; the other pair of cells contained DNA from bacteria and
from other viruses, which the investigators concluded may have
been consumed by the picobiliphytes. That so much information
can be pulled from a trio of cells floating in seawater shows the
power of the emerging molecular technologies as well as how
much we have to learn about the ecology of the oceans.

We have only begun to glimpse the scale of genetic diversity
among planktonic microbes, and have little idea how genomic
variation is mapped upon the three-dimensional mosaic of environmental conditions in the seas. Some microbes are more fas-
tidious than others: one alga may be happy as long as it gets some
sun; another may be restricted to a particular height in the water
column that delivers the perfect temperature and mix of nutrients.[31] Motile protists have some control over their situation;
others reproduce when conditions permit, and then sink into
colder water where they expire. Sampling the water at different
depths reveals highly stratified ecosystems. The size of planktonic
microbes favors their slow descent, and some possess buoyancy
mechanisms that keep them in the sunlight for longer than they

would otherwise enjoy. The glass frustules of diatoms are heavier than seawater, but the regulation of the ionic content of the cytoplasm may go some way toward compensating for the dense shell.[32] Some bacteria and archaea form gas vesicles that keep them floating.[33] The cloud of photosynthetic bacteria in the water column supports predatory protists that make their living as grazers. These organisms have evolved a range of feeding behaviors, but it is impossible to be certain about which ones are used most frequently because we cannot see predation by unicells with the ease that we watch a sea lion eating a mackerel. There are some underwater cameras with sufficient magnification to watch larger components of the plankton—krill and chains of diatoms can be seen—but none can follow an individual cell as it swallows an alga.[34] The problem is compounded by the fact that so few marine microbes have been cultured in the lab. The best we can do is to study a handful of model organisms and make educated guesses about the escapades of their covert relatives.

One of these model predators is a dinoflagellate called *Oxyrrhis marina*, whose lab diet includes bacteria, algal cells, and small flagellates like itself.[35] It will also attack much larger amphipod crustaceans when they become vulnerable during molting. *Oxyrrhis* is an aggressive predator that even turns to cannibalism when foreign cells are unavailable. As it swims in search of prey, *Oxyrrhis* alters the pitch of its helical path according to the density of other cells, moving though a looser spiral as cells disperse, tightening the gyre when it encounters a cluster. Lacking the bug eye of the warnowiid dinoflagellates, its swimming behavior is informed by chemical cues released by the prey. It is like a lioness sniffing the breeze. Even in the lab it can be difficult to see what's happening when *Oxyrrhis* moves in for the kill because its swimming speed of almost one millimeter per second makes it tricky to follow under the microscope. But as it homes in on its victim the dinoflagellate swims in narrow circles, makes contact, fires trichocysts into the prey, binds to its surface, and engulfs its meal by phagocytosis. The horror is over in 15 seconds. The brutality of the water column makes a Sam Peckinpah movie seem like a fairy tale.

The highest concentrations of cells tend to develop in the surface water where photosynthetic activity is greatest. Deeper in the water column, 100 meters or below, where the sea is cold and dark, archaea become the principal form of life.[36] In the first years after their recognition as a separate entity from the bacteria, these prokaryotes were considered to be restricted to extreme environments, the things that could survive in hot springs and hypersaline pools. Subsequent research found the archaea in freshwater lakes, soils, and throughout the seas. Very little is known about the metabolism of the pelagic archaea, but many appear to thrive by a combination of food capture and food manufacture (heterotrophy and autotrophy). Organic matter settling from the sunlit waters is the major fuel source for the cold-water ecosystem, and archaea supplement this diet though the chemistry of *nitrification* powering their metabolism with electrons stripped from dissolved ammonia.[37]

No matter where they live, all of the cells in the ocean are prey to viruses. Little attention was paid to marine viruses until 1989, when Norwegian researchers counted viral particles in water samples using an electron microscope.[38] In a brief paper in *Nature* they reported that a single milliliter of North Atlantic water contained 15 million viruses. Earlier studies had grossly underestimated virus concentrations using a standard microbiological technique in which viruses were quantified by their ability to kill cultured bacteria. A *Prochlorococcus* cell floats with 100,000 of its cyanobacterial clones in a single milliliter of seawater; 10 million viruses called *bacteriophages* bob alongside, primed for their destruction. Recent experiments suggest that a subset of these viruses, called *pelagiphages*, can reproduce in even higher numbers by attacking the SAR11 group of heterotrophic bacteria including *Pelagibacter ubique*.[39] It is estimated that marine viruses slaughter 20 to 40 percent of all of the bacteria in the ocean every day.[40] These mind-boggling numbers of cells and viruses seem to suggest that the sea is a soup of DNA, but, even at these concentrations, the productive sunlit ocean surface is a thin salty broth. A simplified reading of the recipe for making the uppermost part of the water column recommends adding ten parts bacterium and one part virus to one billion parts ocean.[41]

Bacteriophages that infect *Prochlorococcus* and other cyanobacteria are called cyanophages. These particles, or *molecular organisms* (see Preface), have a proteinaceous structure that is likened, for obvious reasons, to the Apollo Lunar Module. The genetic material is packaged in an icosahedral head that connects to a tubular tail that is capped with a base plate surrounded by thin spider-leg fibers (Figure 15). The fibers flex when they make molecular contact with the surface of a bacterium—the molecular analog of a ballet dancer's grand plié—pulling the base plate onto the host wall. The binding of the plate to the wall causes a contraction of the tail, which drives the viral DNA into the bacterium. The viral DNA is translated into viral proteins using the molecular machinery in the host, and within a few minutes the cyanobacterial cell has manufactured 100 or so new phages that burst through its wall into the seawater.

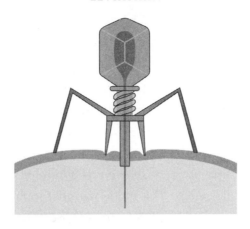

FIGURE 15   Bacteriophage injecting its DNA into a host cell.

Given the reliance of the cyanophages upon the cellular machinery of their hosts, it isn't surprising that they contain a few hundred genes rather than the few thousand genes that constitute the cyanobacterial genome. There is, however, evidence of considerable sophistication in the function of some of the viral genes. For example, cyanophage genomes contain genes that support the photosynthetic activity of their hosts.[42] These include the instructions for making a protein that is vital to the water-splitting reaction of photosynthesis that liberates oxygen. The same protein is also encoded in the host genome, but its incorporation in the cyanophage is essential because the virus disables the host genes to ensure that all of the energy of the doomed cell is funneled into making more virus proteins. This is the microbial equivalent of the psychopath who forces his victim to cook him a full breakfast before ending the festivities. The variety of marine viruses is staggering: one study identified 7,000 types of virus in 200 liters of Californian seawater.[43] There are numerous groups of phages that infect bacteria, plus marnaviruses that attack photosynthetic algae, rhabdoviruses that infect shrimp, herpes-viruses that cause respiratory disease in seals, poxviruses that cause skin lesions in whales, and many others besides. They are, by far, the most numerous nucleic-acid containing entities—"organisms"—in the sea. The coccolithophorids and diatoms that affect the global climate are plagued by their own viruses, and the resulting mortality augments the carbon dioxide pumped back into the atmosphere through other causes of pro-tistan death and decay. The model coccolithophorid, *Emiliania huxleyi*, is infected by giant DNA viruses, one of which contains a gene involved in ceramide synthesis.[44] Ceramide is a lipid ingre-dient in anti-aging creams and investigators postulate that its pro-motion by the marine virus may keep its infected alga going long enough to complete viral assembly.

The discovery of a new kind of giant DNA virus in 2003 vital-ized research on marine virology, and pitted investigators against one another as they argued about the nature of these infectious particles as living entities and reached different conclusions about their evolutionary origins.[45] Ironically, these viruses had been overlooked in seawater samples because they are so big. Most vir-uses are much smaller than cells and can be separated from them using a filter with a pore size of 200 nanometers: the cells collect on the filter and the viral particles flow through. Looking for the giant viruses in fluid strained through a conventional filter would be as futile as trying to wash elephants through a colander. The largest DNA virus described at the time of writing goes by the Latin name, *Megavirus chilensis*.[46] It is almost 700 nanometers in diameter, which is larger than many kinds of bacteria. Its genome is equally impressive, encoding 1,120 proteins including enzymes involved in DNA replication, manipulation, and repair and the synthesis, folding, and chemical modification of proteins. Many of these genes have an inherently non-viral character based on our earlier notions of viral simplicity and non-living nature. This monster was isolated from water collected off the Chilean coast using a freshwater amoeba as bait. The natural marine host of the megavirus isn't known, but is likely to be one or more of the floating protists: there are, after all, plenty of cells in the sea.

Besides killing cellular organisms, viruses operate as genetic re-modelers, incorporating their genes into the genomes of prokary-otes and eukaryotes and acting as couriers for gene transfer between organisms. Through these mechanisms viruses accel-erate the otherwise slothful pace of genetic change through mu-tation and sex. Viruses sabotage the tree of life envisioned by Darwin in which steady modification leads to the proliferation of new branches and demise of older ones, creating an ordered pat-tern of vertical evolutionary descent.

This nineteenth-century model survived the herds of biologists of my generation who felt it sufficient to remodel the phylogen-etic trees of their favorite plants and animals using single-gene comparisons, then analyses using two or more highly conserved genes. And, with few surprises, modern genetics has supported the tree as a model for the evolution of large multicellular organ-isms, and bolstered a robust classification according to the Lin-nean hierarchy of species, genus, family, order, and class.

Disorder is apparent, however, in the vast microscopic ma-jority of life, whose hereditary branches are interconnected hori-zontally via viral interactions, parasitism, mutualistic symbioses, and endosymbiosis. These considerations, coupled with a pen-chant for drama, caused Didier Raoult, the man who discovered giant DNA viruses, to proclaim, there is "no such thing as the tree of life."[47] Creationists shouldn't start popping champagne corks at this news: Darwin captured the process perfectly—life is what works and reproduces—it's just that our language needs to catch up with all that we have learned about the nature and flow of hereditary information since 1859.

Quantum mechanics changed the way that physicists viewed the universe in the twentieth century, relegating experiments with balls rolling on inclined planes to the elementary textbooks and embracing the reality of the very small and the very indeterminate. Biology might have undergone a similar revolution with the invention of the microscope, but we got lost somehow, sidetracked by our unaided eyes and the obviousness of big organisms. For marine biology this meant an attraction to sharks and sea lions and whales. Think about the television documentaries you have seen about the sea. Jacques Cousteau was a favorite of my childhood. He would end his documentaries with phrases that had my sister and me in stitches: "So clumsy on land, so agile in the water" (walrus slips into the water), "Monster of the deep, we salute you!" (Humpback whale disappears into the darkness), and so on. (These are quotes based on my memory of his strong French accent, so consider them as paraphrases.) David Attenborough offered a less disquieting voice, and the Australian guy who wore shorts was a favorite of more recent generations until he was killed by a stingray. The message never changed: the monsters were always the stars.

Returning to my sushi bar, it might be argued that this thought experiment showed a modicum of insensitivity to the loveliest of creatures. Imaginary knives were sharpened for this exercise, I would argue, to make a point. Whales are significant to whales, and their existence is certainly important to the sense of well-being for most people aware that such animals exist. There is an untested assumption, however, that animals as big as whales must matter in a wider ecological sense. One way in which cetaceans have been considered significant participants in ocean productivity is through the contribution made to the organic matter in the seas when they die. The funeral descent of whale carcasses to the sea floor supports rich communities of detritus-loving fish and invertebrates, and has the additional effect of removing carbon dioxide from the atmosphere for good: carbon that reaches Davy Jones's locker stays there forever.

If the stocks of great whales were restored to pre-industrial levels—340,000 blue whales, for example, versus the 5,000 beasts of today's oceans—annual deaths would remove 160,000 tons of carbon.[48] Based on average carbon footprints, the annual mortality of restored whale stocks would balance the climate-destroying power of 30,000 Americans, 60,000 Russians, or 600,000 Tajikistanis.[49] This is significant in the sense that it would match the amount of carbon sequestered by an 800 hectare forest, but is imperceptible when compared with the 2–6 *billion* tons of carbon that the snowfall of dead microbes delivers to the sea floor.[50] If we want to understand the functioning of marine ecosystems, and explore the past and future climate, almost everything that matters is microscopic.

It is often said that we know more about the moon than the ocean depths. Contesting one topic of inquiry against another for the title of "Subject of Greatest Human Ignorance" may be useful as an appeal for funding, but it is, otherwise, a futile exercise: there are so many rivals. And this specific comparison does not withstand the slightest critical thinking. We have determined that cheese is not a signal constituent of lunar geology, yet we lack a wholly satisfying picture of our satellite's composition and origins. There is a lot to learn about the moon. We have many queries about the functioning of the marine majority of the planet's surface too, but we have been discovering a lot about the biology of the sea lately by ignoring all the fish.

4

Dirt and Water

Beast of the field, and over all the Earth,
And every creeping thing that creeps the ground.
—*Milton, Paradise Lost, Book VII*

Outside the village of Fairhaven, Ohio, a settlement that last enjoyed prosperity in the nineteenth century, huge trees overhang a farm road creating a natural tunnel of dark green foliage. This is a magical place for me, reminiscent of the steep-sided roads cut through the chalk of the beech hangers in the Chiltern Hills of my birthplace. Driving through the Fairhaven tunnel is cause for a deep breath: this is beautiful says my brain, and I nod in agreement before rising into the treeless farmland on the other side.

It isn't surprising that we pay attention to trees. Nor is there any intellectual discredit in our fascination with big animals. These are evolutionary imperatives, hard-wired responses to the most obvious, useful, and dangerous organisms in our environment. Concentration is needed to comprehend smaller things: a wasp sting redirects the victim's gaze in an instant, but the truly

Lilliputian requires a conscious decision to find a magnifying lens. Microscopes allow closer scrutiny, and the modern investigator can call upon molecular biology to examine nature at the level of raw information. Appreciation of the very small demands imagination too, a faculty that we ignore too often in the process of scientific inquiry.

Charles Darwin made an imaginative hop, shy of a leap, toward life unseen in his last book, *The Formation of Vegetable Mould through the Action of Earthworms*, published six months before his departure in 1881.[1] He had become interested in worms after his Beagle voyage, and the book described the fruits of long-running experiments in his garden at Down House. To study how worms drew leaves into their burrows he scattered narrow triangles of writing paper "rubbed with raw fat" over the ground. Other observations allowed him to estimate how fast soil displacement by worms buried broken chalk and coal cinders. He concluded that the common European earthworm, *Lumbricus terrestris*, transforms leaf material and other organic matter in a hectare of soil into more than 26 tons of fecal casts per year.[2] Darwin wrote, "It may be doubted whether there are many other animals which have played so important a part in the history of the world, as have these lowly creatures," and compared the global significance of earthworms with tropical corals.[3] Some of Darwin's contemporaries thought this nonsensical. How could such little things do so much? Like his evocation of slow evolutionary change in *The Origin*, Darwin said that his conclusions about worms being nature's plows required his readers to "sum up the effects of a continually recurrent cause."

Obstacles to accepting the logic of evolution and the importance of earthworms are similar to the challenge in comprehending the overwhelming consequence of microbes throughout biology. There is a queerness to all three facts of life. A pinch of soil may seem inert, but contain one billion bacteria and tens of millions of fungi and protists. Ten thousand bacteria could be squeezed inside the period at the end of this sentence, one gram of pure bacteria contains two trillion cells, and so on. Without a microscope and a little imagination these numbers are incomprehensible. The numbers of astronomy are equally perplexing, perhaps more so. Even with a telescope and opportunity for contemplation, the numbers of stars in our galaxy and the tremendous distances between them remain unfathomable. It is very difficult to translate both sciences into digestible facts. Interstellar distances are very big and stars are exceedingly numerous; microorganisms are very small and exceedingly numerous.

Before leaving Mr. Darwin, I'd like to recall his "entangled bank" described in the closing paragraph of *The Origin*:

> clothed with many plants of many kinds, with birds singing on the bushes, with various insects flitting about, and with worms crawling through the damp earth, and to reflect that these elaborately constructed forms, so different from each other, and dependent on each other in so complex a manner.

The steep green banks beneath the hedgerows along the roads around Down House fit this description, and some Darwin scholars think that he was inspired by a particular spot in a local wood that is protected as a nature reserve today. He could have chosen a rainforest for the finale of his book, but the quotidian was a more powerful symbol for his readers: "And all amid them stood the tree of life." Beautiful stuff, but as an emblem of biological diversity, the entangled bank is no more than an echo from the beginnings of modern biology.[4] Life on land isn't quite what it seems.

Plants originated one billion years ago, plus or minus 100 million years, in the ferment of intermingling cells in the Precambrian sea that birthed the ancestors of the supergroupings of eukaryotes. Plants are *embryophytes*, clustered with the red algae, green algae, and charophytes to form the archaeplastid supergroup of eukaryotes. The archaeplastids are united by genetic similarities that suggest that a protist resembling a kind of photosynthetic alga called a *glaucophyte* was their common ancestor.[5] This progenitor was formed when a eukaryote that fed by phagocytosis, like an amoeba, absorbed a cyanobacterium. Rather than digesting the bacterium, predator and prey in this providential interaction sustained one another and reproduced with sufficient synchrony that daughter cells derived from both partners remained paired. As hundreds of thousands of years passed, then millions, most of the genes from the cyanobacterium were transferred to the nucleus of the eukaryote. The time lag between the original acquisition and the development of an irreversible union—with neither symbiont being able to function apart—isn't known, but at some point they cooperated as a mélange of interacting parts whose independent origins vanished into their genomes.

From this cell type came the red algae, the green algae, and the plants. If this seems a plausible idea rather than a fact, this is only because I have not invested enough words in detailing the genetic data. There is, in truth, no other sensible explanation for the available evidence. For instance, one fifth of the genes in the nuclei of some plants have an obvious cyanobacterial origin, and the chloroplasts of all of the achaeplastids have their own circular chromosomes that encode a reduced subset of the genes found in free-living cyanobacteria. Endosymbiosis is a mainspring of modern biology.[6]

This early endosymbiosis resulting in a glaucophyte, or an alga that looked a lot like a glaucophyte, is likely to have formed in a marine environment. The subsequent transformation of marine biology into the pioneering forms of terrestrial biology is the subject of educated guesswork: we know that this happened, but we don't know how it happened. Earth is 4.6 billion years old and life was sparked within the first billion years of its solar birth. The primordial marine cells may have resembled prokaryotes that populate contemporary ecosystems around hydrothermal vents in the deep ocean. These included hydrogen-fueled bacteria and other kinds of chemotrophs that made sugar in the dark. Eukaryotes were engineered from a prokaryote ancestor, or via the fusion of a pair of prokaryotes, but there was nothing multicellular for a long time and nothing at all on land for even longer. Ultraviolet radiation streaming from the star prevented sunbathing.

This appalling situation began to change 2.5 billion years ago when the atmosphere filled with a highly reactive and corrosive gas produced by cyanobacteria that had developed the trick of oxygenic photosynthesis. With this innovation, anaerobes perished in the sunlit ocean as oxygen bubbled from mineralized pillars of cyanobacteria called *stromatolites*. Above the water, high in the atmosphere, ultraviolet radiation created its own stratospheric shield by splitting O_2 molecules into separate oxygen atoms that combined with intact oxygen molecules to form ozone (O_3). This novelty allowed life on land to launch itself as a sun-baked crème brûlée of prokaryotes and algae. With a time lag of a few hundred million years, primeval land plants plastered themselves over this crust, or paleosoil, in the Ordovician.[7]

Green algae formed the evolutionary link between glauco-phytes and plants. These photosynthetic protists include swimming unicells propelled through the water with two or four flagella, filaments, colonial species whose cells are joined into iridescent orbs, networks of tubular cells that look like inflatable fun houses, an alga called the sea lettuce, and beautiful siphona-ceous algae with giant cells made crunchy by calcium deposition in their walls.[8] The most complex of these protists are nested within a subgroup called the streptophyte algae that includes the filamentous alga *Spirogyra*, *Coleochaete* that produces tiny whis-kered discs, and the Charales or stoneworts. The first land plants arose from somewhere within this cluster of aquatic organisms. They resembled liverworts and mosses and possessed few adaptations for exposing themselves above damp soil. If we fast-forward to the flowering plants, that grew first as weeds 140 million years ago, adaptations to life above the ground are abun-dant.[9] These fully terrestrial archaeplastids had vascular tissue connecting their leaves to their root systems; their leaves were sun-screened and kept hydrated by a waxy cuticle; and their sperm were delivered to their eggs via pollen tubes, emancipating plant sex from its aquatic trappings.

Today, the greenery is populated by 250,000 to 400,000 spe-cies of flowering plants and 40,000 species of other kinds of plants, including other seed plants, ferns, horsetails, mosses, and liverworts. These are big numbers, reflecting the profusion of botanists dedicated to naming plants since the eighteenth century.[10] Yet there is a uniformity to all of the plants, which is impossible to appreciate at first glance. The range of leaf shapes and sizes and the brilliance of floral forms are powerful distrac-tions from the fact that all plants work in the same way and add no more than a splotch to the astonishing breadth of biological diversity when viewed in relation to the supergroup arrangement of life. The sameness of plants is evident in their way of life. Every green plant makes its living by virtue of chlorophyll-containing chloroplasts, and every one of these organelles has been inherited through a single kind of alga, the prosperous glaucophyte that co-opted a cyanobacterium as its onboard confectioner one billion years ago. If you understand how a liverwort makes sugar, you know how every other plant works. As far as the shrubbery stretches, there is nothing new under the sun: all of the mechanisms evolved a very long time ago at sea, and every plant is a contrivance that propagates an encyclopedia of ancient prokaryote genes. Plant chloroplasts are cyanobacteria surrounded by a membrane. Trees are contrivances for lofting bacteria skyward to give them a clear view of a yellow dwarf.[11]

The repetitiveness of the greenery—the monotony of botany—notwithstanding, all of us appreciate plant taxonomy as a matter of survival. Differences in the look, taste, and smell of fruits and vegetables are controlled by a relatively small numbers of genes, but we are very skilled at distinguishing between these novelties. While our knack for telling plants apart has tremendous adaptive value, it is a formidable obstacle to comprehending the vastness of life. Attuned to plants, we overrate their diversity, and fail to appreciate the smaller forms of life. A pair of protists may be separated by greater genetic differences than a moss and an orchid, yet look exactly the same under the microscope. This is the reason that we have tended to hide most of life under the *Amoeba* or hairy bacterium at the base of popular depictions of the tree of life. Genetic data have transformed the picture, but these particulars of biology are known to few besides the scientists engaged in the research.

A second way in which we misread life is through the custom of treating plants, and other multicellular organisms, as autonomous entities. A rose is partly a rose, but mostly it is lots of other things. Its roots, leaves, stem, thorns, and flowers are caked with microorganisms, its inner anatomy is riddled with other life forms, and the whole magnificent botanical enterprise of the rose is sustained by a medley of soil bacteria and fungi. A rose does not exist outside this teeming collective farm. The bank is more tangled than Darwin ever imagined.

* * *

Logic suggests that Soil Microbiologist is a safer profession than Marine Microbiologist, the latter being prone to slipping over the side of research vessels while hauling plankton nets from the depths. Health hazards aside, it is just as difficult to figure out what's living in the dirt and how they're doing it as it is to unravel the biology of the open ocean archaea. The incredible diversity of microbes in both environments escaped investigators for a long time. Much of the fault for this colossal omission lay in misplaced trust in the orthodox methods of microbiological research. If a fragment of soil is spread on an agar plate and incubated overnight, the morning will find the jelly webbed with glassy filaments of *Penicillum*, *Aspergillus*, and other fungi, and pock-marked with shiny colonies of common yeasts and bacteria. The harvest can be altered by adjusting the quantity of sugars and other ingredients mixed with the agar to advantage microorganisms that are otherwise obscured by the plebian. Fungal growth is amplified by the provision of antibacterial antibiotics that halt non-resistant bacteria; the addition of antifungal agents reverses the experiment.

Many researchers have spent their entire careers working with these methods, cataloguing the fungi and the bacteria cultured from soils, and their efforts are evidenced in a voluminous literature. They missed a great deal. For every one of the tens of thousands of microbial species documented in this meticulous fashion there are more than 100 kinds of soil fungi and bacteria that have never been seen. The species that prosper on agar represent less than 0.5 percent of the life in the soil; most microbes have thirsts that we have failed to slake in the laboratory. Classical microbiologists acted with vanity comparable to the missionaries who dangled hand mirrors from trees in the hope of persuading the savages into the open where they could be civilized with the gospels. Neither explorer learned much about the undisturbed territory.

The development of techniques for amplifying genetic sequences without needing to see anything on a culture dish is changing our picture of the soil as much as it has transformed microbial oceanography. Soils harbor even greater microbial riches than the oceans. Part of the reason for this is that the structure and chemistry of soil varies so much. The composition of the communities of marine microbes varies according to their vertical position in the water column and, horizontally, according to temperature, local nutrient conditions, and other variables. Even so, there can be homogeneity to the physical habitat over thousands of square kilometers, with outpourings of nutrients from a river delta or volcanic eruption sustaining clonal phytoplankton blooms visible from space.

Soil is very different, with a three-dimensional structure and chemistry that vary over distances ranging from the microscopic to the geographic. The construction of soil is determined by the proportions of sand grains, tiny fragments of silt, and clay particles that are smaller even than bacteria. These are the mineral constituents of soil. Soils dominated by sand tend to be well aerated but less fertile than those with a greater proportion of silt and clay. Clay has a crystalline structure that attracts ions from percolating water and can exchange them with plant roots. Because clay particles are so small, they offer an enormous surface area for these chemical reactions: a cubic meter of clay has a surface area of 6,000 square meters. Soils support plant growth, plants support animals, and decomposing plant and animal tissues, together with animal feces, enrich the soil.

One gram of rich forest soil contains an estimated 100 million prokaryotes.[12] Measurements of soil biodiversity have been based on the metagenomic techniques used for seawater samples and from a cruder method involving the melting and reassembly of DNA strands.[13] The second procedure treats all of the bacterial chromosomes in a sample as a single unit. When this bulk DNA is heated, the strands of the double helix separate; when they are cooled, they recombine at a rate dependent upon their size. The kinetics of these *reannealing* reactions reflect the size of the collective soil genome and afford estimates of the number of different genomes in the sample. On average, 1,000 kinds of DNA will behave like a monster chromosome that is 1,000 times bigger than each of its components. Early experiments with this method suggested that a gram of soil contained many thousands of different species, or genetically distinct strains. More recent research boosted the estimated diversity to as many as one million taxa per gram, signifying that most kinds of soil bacteria and archaea are very scarce and are swamped by a governing majority.[14]

The complexity of soil biology and the obstacles to understanding its microbial inhabitants has prompted a consortium of researchers to embark upon an international sequencing project to unravel the "TerraGenome."[15] The scale of the endeavor is fantastic, dwarfing the technical challenge that faced researchers who completed the task of sequencing the human genome. Although sequencing technology is improving rapidly, the projected cost of the project is enormous, but the prospects for agricultural and biotechnological innovations resulting from a better understanding of the way that soil works are considerable.

On the other hand, there are a number of objections to the work.[16] Soil structure and chemistry are hugely variable: compare, for example, a desert soil with low moisture and little organic matter, with a rich "andisol" derived from volcanic ash that supports rice cultivation. This limits the usefulness of a metagenomic analysis of any "representative" soil sample. Every soil is a mosaic of short-lived microhabitats whose occupants are mashed together by bulk DNA extraction.

Finally, genomic analysis reveals next to nothing about interactions between particular microbes and larger soil organisms. A novel assembly of genes reveals that a particular bacterium exists in a soil sample, but doesn't tell us what it does, where precisely it operates, or with whom it associates. The genes of this cryptic organism may flag it as a methane-generating anaerobe, but the cells could be ticking over, rather than farting with gay abandon, and contributing little to the chemical balance of the soil. At the time of sampling, it might have been waiting for a dip in oxygen levels that would permit its return to normal flatulent programming. In the pre-genomic infancy of soil biology, when experimentation was limited to microbes that were amenable to lab culture, we had learned about a few of the dominant collabor-ations in the "soil microbiome." But this was a stunted view of dirt that overlooked almost everything that lived in it. Genetic discovery has transformed dirt science from a seemingly tedious study of soil chemistry into an epic inquiry into the grandeur of life.

Bacteria and archaea employ a variety of energy sources to make a living in the soil. Sunlight powers cyanobacterial photo-synthesis and releases oxygen when electrons are stripped from water. Photosynthesis is also accomplished in anoxic environ-ments by purple and green bacteria that take their electrons from hydrogen sulfide, elemental sulfur, and other substances. Archaea don't do photosynthesis according to these methods, but some of them use sunlight as a supplementary power source using a pro-tein called bacteriorhodopsin that is related to the rhodopsin pigment in our retina. Bacteriorhodopsin is concentrated in the archaeal cells in crystalline patches called "purple membrane." *Chemolithotrophic* prokaryotes feed themselves by oxidizing inor-ganic compounds, using the reducing power of electrons reaped from these substrates to produce ATP. Hydrogen gas is a popular fuel and other bacteria oxidize sulfur (producing sulfuric acid), ferrous iron, ammonia, and nitrite. Heterotrophic bacteria and archaea derive their nutrients from other organisms, sometimes forming supportive relationships with their food sources, also acting as parasites, and participating in the frenzied decompos-ition of dead soil organisms of every stripe. Many prokaryotes require oxygen for their decomposing facility, some are content to ferment without it. *Methylotrophs* save themselves the effort of degrading complex materials and consume simple organic mol-ecules including methane, methanol, and formate. This paragraph covers a good deal of the diversity of soil prokaryotes, but is not an exhaustive list of lifestyles.[17]

Microbes have dominion over the planet's nutrient cycles and arbitrate the distribution and abundance of plants and animals. Nitrogen is the most plentiful element in the atmosphere, and every living thing needs it to assemble its nucleic acids and proteins. Only bacteria can pull nitrogen gas out of thin air and cast it into the stuff of life. The conversion of nitrogen to ammonium is catalyzed by an enzyme called *nitrogenase*. Nitrogenase is poisoned by oxygen and the bacteria that invest in its synthesis operate in anoxic habitats or protect the enzyme by mopping up all of the oxygen in the vicinity of the protein. Filamentous cyanobacteria that fix nitrogen create the necessary environment inside triple-walled cells called heterocysts that exclude oxygen. Nitrogen-fixing bacteria form symbioses with plants, including the well-known root nodules of legumes in which an initial infection is transformed into a stable mutualism where the bacteria are fed by the plant in exchange for ammonium. The plant supplies a trickle of oxygen to support the respiration of its bacteria, without poisoning the nitrogenase, using leghemoglobin, a red pigment whose structure resembles our own hemoglobin. Soil bacteria drive every other reaction in the nitrogen cycle, with nitrifying species converting ammonium to nitrite and nitrite to nitrate, and denitrifying kinds driving the reactions in the opposite direction and returning nitrogen gas to the atmosphere.

Bacteria use nitrifying reactions as a source of energized electrons; denitrifying bacteria use nitrate in the same way that we use oxygen to soak up electrons exhausted by our metabolism. These free-living soil bacteria, as well as those that capture other nutrients, bloom in the highest concentrations in the vicinity of roots. This soil territory is called the *rhizosphere*, and is a special-ized habitat of marvelous complexity. The metabolic activities of all of the rhizosphere bacteria and other microbes show ex-quisite sensitivity to soil chemistry. Experimental manipulation of the gaseous environs of these communities suggests that in-creasing concentrations of atmospheric carbon dioxide will upset the fine-tuned recycling of nitrogen and other elements and intensify the production of an assortment of potent green-house gases.[18] This is a forceful example of the significance of environmental microbiology in evaluating the merits of the scorched earth policy promoted by the coal, oil, and gas industries.

The nitrogen cycle is one of the core topics in ecology courses and illustrates the dependence of plants upon bacteria, agricul-ture upon bacteria, and us upon bacteria. Thirty years ago, I became fascinated by a lichen that contributed to this chemistry when the new chair of my undergraduate botany department, Tony Walsby, participated in a competition judged by the stu-dents. The botany professors were allotted a few minutes each to argue, on the behalf of anything other than an animal species— for which the zoology department was accountable—for the title of "The Greatest Organism."

Walsby is an expert on gas vesicles, proteinaceous capsules that keep cyanobacteria and other kinds of aquatic prokaryote afloat. He explained that *Peltigera canina*, the dog lichen, was a super-organism formed by a cooperative trio of species. A fungus provided the framework for the symbiosis, acting like a sponge in wet conditions, creating a sheltered habitat for the other microbes; a photosynthetic green alga absorbed carbon dioxide, manufactured glucose, and converted this into sugar alcohols that it shared with the fungus, and a cyanobacterium carried out photosynthesis and also fixed nitrogen from the air. As long as the sun shone and the lichen was washed with water carrying some dissolved minerals, *Peltigera* could meet all of its needs. Walsby's argument was brilliant: no other organism shows such self-sufficiency. At the conclusion of the contest, he was disqualified for sponsoring a symbiosis rather than a single organism. My research mentor, the mycologist Mike Madelin, won handily by reading from Virgil's *Georgics* and advocating *Vitis vinifera*, the grape vine.

Most of the 15,000 species of lichenized fungi partner with a green alga alone, less than 10 percent contain cyanobacteria, and *Peltigera canina* is one of a few hundred tripartite symbioses.[19] The green algal symbionts come from the genera *Trebouxia* and *Trentepohlia*, and *Nostoc* is the most common cyanobacterium in lichen symbioses. *Trebouxia* seems to be stuck with life inside lichens, whereas *Trentepohlia* and *Nostoc* have other options and can support themselves without a fungus.

The connection between the fungus and its photosynthetic partners is very intimate. The green alga is penetrated by extensions from the fungal filaments called *haustoria*, which resemble the feeding structures formed by parasitic fungi, like rusts, and their plant hosts.[20] The fungus squeezes its cyanobacteria in a different way, sending slender protrusions through their gelatinous sheaths to finger the filaments within. Experiments suggest that the fungus taps 90 percent of the sugars produced by their partners, which raises the same question posed by the coral symbiosis with dinoflagellate algae: is this mutualism or parasitism?

This question engaged nineteenth-century botanists who were first to recognize that lichens were more than single organisms.

At that time, consensus fell upon the idea that lichen fungi "enslaved" their algae, but Anton de Bary, one of the early investigators, was wary of making fundamental distinctions between parasitism and mutualism, recognizing "every conceivable gradation" among interactions between symbionts.[21] There is a tendency to ignore De Bary's subtle treatment of interspecies relationships, and we remain vulnerable to viewing biological interactions as good or bad, mutualistic or parasitic. The good symbiosis is exemplified by the clownfish that feeds on the anemone's pests, feeds the fortunate anemone with its feces, and is afforded protection by the stingers of its ever thankful partner; the bad by the filarial worm infected by *Wolbachia* bacteria that causes elephantiasis and can make an ill-starred gentleman's testicles swell to the size of basketballs. In many other relationships, the energy flow between participants is unclear, and we have no way of measuring the reproductive benefit of the union relative to an independent living. For this reason, the use of De Bary's broad concept of symbiosis allows us to sidestep the uncertainty of categorizing an organism as a mutualist or parasite.

Another way to think of the lichen is to view the symbiosis as a form of algal husbandry, an example of farming by a fungus. And other fungi are, in turn, farmed by tropical ants and termites.[22] Mycorrhizal symbioses between fungi and plants are another example of complex interactions that are difficult to categorize. Associations between fungi and plant root systems were discovered around the same time that lichens were recognized as symbioses. Connections between mushroom-forming fungi and trees are evident from their stubby roots, which are surrounded by a mantle of fungal cells and penetrated by their filaments. These ectomycorrhizal relationships are not widespread in terms of the diversity of plant species involved, but their support of the dominant trees in temperate and boreal forests means that the biosphere would collapse without them. Other kinds of mycorrhizal fungi penetrate the roots of 80 to 90 percent of every other plant species, and these cross-kingdom associations are as old as plants themselves. Mycorrhizal fungi preserved as 460-million year old fossils of spores probably infected ancestral liverworts or mosses, given that no plants had realized greater anatomical extravagance then.[23] Similar fungi associate with liverworts today, forming coils and finely branched arbuscules inside cells of the slivery green thalli of these simple plants. In this position, the fungus is placed perfectly to receive synthetic goodies from the liverwort's chloroplasts, and, in return, it scores out through a collective 400 meters of wet soil drawing upon a vast field of dissolved nutrients to support this interdisciplinary marriage.[24]

Soil fungi are not limited to the mycorrhizal species, and portions of my previous mycological books have been invested in their wood-degrading, insect-infecting, and other-fungus-infecting lifestyles. Besides writing, I have taught undergraduate courses on various aspects of mycology at my university for almost 20 years, and I'm struck, each year, by how much revision is required in presenting current thinking on the diversity of these microorganisms. Mushrooms produced by colonies of soil basidiomycetes are the most familiar manifestation of the fungi. Along with plant pathogenic rusts and smuts they account for 40 percent of the 72,000 described species of fungi. The ascomycete fungi, including the yeast, *Saccharomyces cerevisiae*, account for a similar proportion of the organisms studied by mycologists. The remaining 20 percent of the fungi is the most enigmatic slice of the group and is where most of the genetic diversity lies.

Frederick Sparrow, a mycologist at the University of Michigan, authored a monograph titled "Aquatic Phycomycetes" in 1943.[25] This monumental work was among the early efforts to impose taxonomic order upon the multitude of fungi that swarm in wet soils and ponds, digesting the deluge of plant and animal parts from above ground, infecting protists, infecting one another too, and creating gorgeous microscopic networks of blobs connected by fluid-filled canals, flasks that belch swimming spores, and branched cells resembling miniature fruit trees.

The lasting impression of Sparrow's monograph for me—and I am aware of my unique sensitivity to these things—is bittersweet. The fact that a good deal of Sparrow's classification schemes have been overturned by subsequent research, including the transfer of some organisms from the fungi, or opisthokont supergrouping, to the stramenopile protists, is of little importance. The greater problem is that so many of Sparrow's microbes are lost to science: the fantastic manifestations of life showcased in the monograph have not been glimpsed in decades. These lost organisms include bizarre-looking fungi like *Megachytrium westonii*, which he discovered on pond weed in Ithaca, New York, and the fabulous *Araiospora* that must be a close relative of the Flying Spaghetti Monster (Figure 16).

Paleontologists enjoy brief fantasies about watching the fledgling pterosaurs leave their nests on their inaugural flights. A computer animation is the closest they will get to the living reptiles. Prospects for satisfying fans of Sparrow's lost world are much brighter. Some of his fungi may have suffered extinction, but most are out there right now, doing the same things they have done for tens or hundreds of millions of years. The reason that we know so little about them is that the tiny subset of humans who

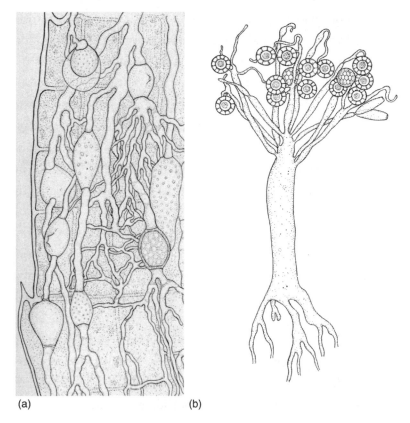

(a)   (b)

FIGURE 16   Aquatic microbes showcased by Sparrow. (a) Colony of chytrid fungus *Megachytrium westonii* (Opisthokonta) within leaf cells of Canadian pond weed. (b) Thallus of water mold *Araiospora pulchra* (Stramenopila).

Source: F. K. Sparrow, *Aquatic Phycomycetes (Exclusive of the Saprolgeniceae and Pythium)* (Ann Arbor, MI: University of Michigan Press, 1943).

study biology must select a tiny subset of life for close inspection, and Sparrow's fungi weren't picked as "model organisms."

Our ignorance—and I use this word in its least pejorative sense—is inevitable, but among these mysterious microbes are unimagined genetic riches, the molecular blueprints of complex symbioses that we may never know. And we learn more about the depth of our naïvety with every gene amplification from nature. Thanks to the devotional work of Sparrow and his students, and their students who kept working on these obscure organisms, we knew a little about chytrids before one of them was linked to the pandemic infection and mass killing of amphibians in the 1990s. In short order, a species in this group of disregarded fungi, called *Batrachochytrium dendrobatidis*, became the object of intensive investigation.[26] Our ignorance about microbes will find us stranded in the future, if something *really* anonymous demands our attention.

Another group of fungi, the *Cryptomycota*, offers an interesting example of an overlooked part of nature. Sparrow included the fungal genus *Rozella* in his 1943 monograph. These microbes produce single-celled blobs that grow inside the cells of other species of fungi and stramenopiles. After they have finished feeding, these thalli are converted into sporangia that release swimming zoospores, or into spiny spores that acts as survival capsules. The original description of *Rozella* was published in the 1870s, but because it could not be cultured outside its natural hosts there had been no experimental investigations on the fungus by the time Sparrow set to work. Some excellent electron microscope studies on the fungus were published in the 1980s, including one report showing *Rozella* infecting another fungus that was infecting a euglenoid alga.[27] Details of its cell structure seemed to situate the beast well outside all of the known groups of fungi. Twenty years later, molecular phylogenetic analysis pointed to the same conclusion: *Rozella* was an outlier that couldn't be squeezed into the available groups.[28]

The end of its obscurity came in 2010, when researchers amplifying genes from a Swiss peat bog scrutinized a set of sequences that didn't nestle within the existing phyla of opisthokonts.[29] These sequences were related, however, to genes sequenced from *Rozella*, and the authors of this study concluded that species of *Rozella* and their relations in the peat bog known only from DNA samples represented a distinctive category of fungi. A second study took things much further, by demonstrating that *Rozella* was representative of a huge assemblage of flagellate fungi that live in soil, freshwater, and marine habitats.[30] In addition to finding these fungi in relatively pristine environments, the researchers fished their DNA from polluted sediments and chlorinated drinking water. They seem to be everywhere, and the spread of their DNA sequences encoding a particular diagnostic gene was greater than researchers have documented across all of the known fungi. Most of them operate, like *Rozella*, as parasites of other microbes; none have been cultured. The investigators called the group the Cryptomycota, suggesting that it should be treated as a new phylum.

This was a sensational discovery. The creation of a new phylum is a big deal because phyla are viewed as very big categories of organism. Phylum Chordata, for example, subsumes animals as disparate as rabbits and sea squirts. There is no objective definition of a phylum, but the general notion is that this category encompasses a wide assemblage of things that can be differentiated from related groups by an assortment of characters. The unearthing of the Cryptomycota is interesting for lots of reasons, not least because it raises the question: what else is out there? The opportunities for the bioprospector are limitless. One can examine a substrate that few have probed, or pick apart the genes of cryptic organisms and look for similar instructions in familiar places. The hopefulness of the quest was posed in a rather unsavory fashion by a fungal taxonomist, R. W. G. Dennis (1910–2003), in a passage of introductory humor in his magnum opus, *British Cup Fungi and Their Allies*:

> It is a fortunate borough whose pavements are not daily fouled by dogs yet little is known of the fungi that colonize this substrate. Nearly half a century ago Crossland observed interesting species of *Rhyparobius* and other [fungal] genera on "canine excrement" but hardly anyone has looked since his day. A rich harvest may well await the man who cares to devote his leisure hours or his declining years to the study of stale dog dung.[31]

This howling lunacy deserves its own grave marker, but is, nevertheless, a statement of microbiological truth. Prospecting for other kinds of microorganism is equally rewarding, and viral discovery is available to anyone acquainted with elementary methods of molecular biology.

Phages, as we saw in the previous chapter, are viruses that infect bacteria. Staggering numbers of these particles—hundreds of millions per gram according to some estimates[32]—infect the abundance of bacteria in soil and connected freshwater ecosystems. Phages are convenient research subjects for students of genomics because they possess tiny genomes, and the "Phage Lab" has become a standard exercise for undergraduates in some universities. Phages are isolated from environmental samples by allowing them to kill a single species of bacterium on culture plates. The microbe sacrificed for these experiments is *Mycobacterium smegmatis*, a harmless relative of the bacteria that causes tuberculosis and leprosy. This fast-growing species was isolated originally from a syphilis ulcer, or chancre, but is unfazed by transfer to agar. A single viral particle can infect a single bacterial cell, reproduce in geometric fashion, invade adjacent bacteria, and so on, until the confluent lawn of the bacterial culture is marred by an empty patch, or plaque. If a sample of soil is diluted, mixed with bacteria, and spread on agar, the phages will dot the culture with plaques after 24 hours of incubation. These plaques may be initiated by more than one phage, and the students engage in further rounds of isolation and bacterial infection to ensure that they have purified single phages. After many more steps in this lengthy procedure, the students purify and sequence the phage DNA, and can submit their sequences to an online database.

Because the students use the same bacterial species as bait, they all isolate the same kinds of phage, but the sequence variations among these catches are endless. Almost every phage discovered by a student is novel. Even when a class samples the same soil repeatedly, new phage sequences keep coming. A key website called the Mycobacteriophage Database allows students to post links to their viral sequences and provide electron micrographic portraits of their discoveries.[33] They are also free to name their viruses. Here's a randomly selected trio of phages: *Backyardigan* is a myovirus isolated by a student from Western Kentucky University from, as you may have guessed, his backyard; discovered at a dog park in Missoula, the siphovirus *Misomaster* would have glad-dened the late R. W. G. Dennis; and, another siphovirus, *Patience*, isolated in Durban, South Africa by a student from the Nelson Mandela School of Medicine, was the 1,000th phage added to the database. There are more than 2,400 phages on the website at the time of writing, all hooked by the same species of bacterium. The unfathomable diversity of soil phages is evident from the fact that each type of bacterium is prey to multiple kinds of phage. Estimates suggest that we have characterized only one millionth of the range of these viruses in soil, and the real variegation may dwarf this judgment.[34]

*   *   *

The microbiology of soil and freshwater ecosystems are closely related. Water is cleansed, and often polluted, by its percolation through soils before draining into a pond, lake, creek, river, or estuary.[35] Bacteria, archaea, protists, and fungi in soil are transferred into these environments, and some of them thrive in free water. Many soil microbes that never leave the soil are aquatic organisms, in the sense that their activity is limited to the water-filled crevices of soils. Exceptions include the microbes whose life cycles are harnessed to plants and terrestrial animals. Mycorrhizal fungi are an obvious example of non-amphibious organisms, but there are examples of related species that specialize in wood decay that thrive on land and in water. The discovery of a fungus that degrades submerged twigs and produces its gilled mushrooms under water is a testament to the physiological flexibility of some ostensibly soil-bound organisms.[36]

Microorganisms in the open water of lakes offer the greatest contrast to microscopic life in soils, with complex populations of algae and non-photosynthetic protists that show comparable vertical stratification to the planktonic microbes of the ocean.[37] Oxygen availability and temperature are major determinants of the structure of these communities. Summer warmth reduces the density of the water in the surface of lakes in temperate climates, and this part of the water column is separated from the colder and denser water below by a boundary called a thermocline. If the lake water isn't disturbed, the oxygen levels below the thermocline plummet, driving the proliferation of anaerobic bacteria and archaea. Later in the year, as the surface waters cool, the different layers are mixed, the anoxic zone is infused with oxygen, and the layering of aerobes over anaerobes is disrupted for a few months.

Similarities in nutrient cycling in the pelagic majority of freshwater and marine environments obscure profound differences in the physiology of the microbial operators. The concentration of sodium chloride in seawater is 2.5 percent, a little less than pepperoni pizza, and almost three times brinier than our blood. The contents of freshwater microbes and cells moving through wet soil are considerably saltier than their surroundings. This osmotic imbalance drives water into the cytoplasm. Bacteria and many other kinds of microbes synthesize a strong cell wall on the surface of their membranes that resists expansion of the cell and pressurizes the cytoplasm. The hydrostatic pressure inside the cell balances the gradient of incoming water, and the turgid cell is in osmotic equilibrium with its surroundings. An unwalled cell like an amoeba has no protection from unremitting water influx, but avoids destruction by collecting water in a contractile vacuole that fills with water (diastole) and empties itself (systole) on the cell surface every few seconds. This pulsating vent allows the amoeba to feed by phagocytosis. The filamentous hyphae of fungi are walled, lack contractile vacuoles, and absorb nutrients dissolved from their food by secreted enzymes. This difference in feeding mechanism is one of the fundamental distinctions between walled and naked eukaryotes. Marine microbes suffer little of the inconvenience of water influx because the saltiness of their cytoplasm is closer to the salinity of their surroundings. Transferred to the sea, a freshwater amoeba shrivels to death; transferred to fresh water, the marine microbe bursts at its seams.

The magnitude of population sizes among terrestrial micro-organisms, and the ease with which their cells may be carried by rivers and through the air, suggests that the marine environment receives a continuous deluge of land-based cells. Reverse transmission to the land from sea foam whipped skyward is also significant. Either way, the passive migrants perish: the saltiness of the sea is as impenetrable a barrier to a freshwater diatom as the glacial purity of an Arctic pool presents to a marine dinoflagellate lofted from a Norwegian fjord. This least forgiving of physiological hurdles has kept large groups of organisms in one place or the other. Even over an evolutionary timespan, few migrants have seeded new groups of microbes in either habitat.[38]

There are, for example, 5,000 to 6,000 species of green algae (archaeplastids) called desmids that govern primary productivity in planktonic communities in rivers and lakes with low nutrient levels. No desmid grows in the ocean. The euglenoid algae (excavates) prefer high nutrient levels and very few of them live in the sea. Representing the reverse case, are the red algae (archaeplastids) and the brown algae (stramenopiles): both groups are important marine protists, but a small number of species of each—less than 2 percent of the reds and only 4 of 2,000 species of brown algae—have colonized freshwater habitats. Radiolarians (rhizarians) are exclusively marine, and very few of the more than 200,000 species of related foraminiferans grow in freshwater habitats.

Exceptions have evolved. Diatoms are an example of a large taxonomic group that bucked this xenophobic tendency and are well represented both in lake and sea. Fossils of marine diatoms predate freshwater fossils by more than 65 million years, suggesting, but not proving, a marine origin for these glass-walled protists.[39]

One diatom expert described the physiological feat of moving from the ocean to a freshwater habitat as "crossing the Rubicon," meaning that there was no going back. He was wrong. Analysis of the genetic relationship among round-celled (centric) diatoms demonstrate that freshwater species have arisen from marine groups *and* vice versa on multiple occasions.[40]

The adaptive mechanisms that allowed marine diatoms to avoid explosion in freshwater and freshwater species from desiccating in the sea are not understood, but then we have established so little about the physiology of diatoms in the first place. Exposure to changes in salinity over an evolutionary timescale may have come about via initial colonization of estuaries by diatoms emigrating seaward, and, conversely, by inundation of salt water into freshwater habitats. Either mechanism would have soothed the shock of pure water for migrants who abandoned the ocean. Diatoms aren't the only microbial landward pioneers. The algae that greened the Ordovician made the same transition, and their descendants filled the African forests and grasslands in which we were so recently born.

5

Air

so stears the prudent Crane
Her annual Voiage, born on Windes; the Aire,
Floats, as they pass, fann'd with unnumber'd plumes:
—*Milton, Paradise Lost, Book VII*

Many of my childhood memories derive from my discovery of the wonders of a decaying apple orchard. It was a place of such stunning quietness. "The orchard's lush ceiling trapped a perpetual fog, and in this water-saturated atmosphere trees and fruits were devoured by bitter rot, black rot, blossom end rot, canker, rust, powdery mildew, rubbery wood, and scab."[1]

It was lighter on the other side of the orchard fence, in my garden, where the air danced with life and the minute seeds of its ruination. Spores from the diseased trees poured from the orchard and were caught in light beams along with the random jazz of other microscopic particulates and more deliberate flurry of fairyflies,[2] fungus gnats, and midges. As a child of remarkable and no doubt damaging introspection, contemplation of the glittering specks suggested that there was more to life than the immediately apparent. Lucretius was similarly rapt by "dust-motes dancing in the sun" in

The Nature of Things. By reasoning that the agitation was driven by "unseen atomic blows," the Roman poet prefigured the discovery of Brownian motion by more than 18 centuries.[3]

In 1933, just before the worst of the Dustbowl Years in the United States, Colonel Charles Lindberg collected air samples using a "sky hook" as he flew between North America and Denmark, taking a circuitous side trip north of the Arctic Circle above Greenland.[4] The sky hook, designed by Lindberg, held interchangeable cartridges that contained oiled microscope slides. Unexposed cartridges were engaged at the free end of the device using a rod, opened to the air, closed after the sampling period, and replaced with a fresh cartridge. Lindberg's wife, Anne, was copilot of their single-engine Lockheed monoplane, christened *Tingmissartoq* after an Eskimo word that meant something along the lines of "big bird." She took the controls when Charles pulled back the forward canopy to exchange the cartridges while the icy air whipped around him.

I mentioned the Dustbowl because there was, understandably, a lot of interest in the weather, atmospheric sciences, and the future of agriculture at that time. Severe drought turned the soils of the American prairies—already wrecked by overgrazing and deep plowing—to dust, and the fortunes of farmers who resisted westward migration were wasted by uncontrollable epidemics of cereal rusts. The Lindberg's sky-hook collections were requested by Fred Meier, an expert on airborne microbes at the U.S. Department of Agriculture, who wanted to understand the spread of the rusts. The results of the collaboration were spectacular: 1 kilometer above Greenland the spores of rusts and other fungi plastered the sticky slides. Thousands of kilometers from the source of any fungal growth on the ground, these pernicious microorganisms were moving around the planet, high in the atmosphere. The atmosphere was brimming with disease.

Meier's research also involved less ambitious flights of military and commercial aircraft, and he flew on the USS Los Angeles, a helium-filled airship, and collected samples of fungi that he cultured back in the lab. He was a charismatic scientist who attracted seemingly limitless support, financial and otherwise, for his research. Samples were obtained for Meier during a record-breaking balloon ascent to an altitude of 22 kilometers in 1935, and Amelia Earhart was collecting for him shortly before her disappearance in 1937. The next year, Meier vanished with the Hawaii Clipper, a Pan American flying boat whose last radio transmission, southeast of Manila, was, "Stand by for one minute as I am having trouble with rain static."[5] Aerobiology lost its greatest advocate and has never recovered. Research on the airborne microbiome has continued, and there have been some fascinating discoveries, but there is a definite sense that this is a scientific backwater, a subject that no longer deserves the attention of the greatest minds. This is a grave mistake.

Those fortunate to live outside zones of conspicuous pollution perceive air as a transparence surrounding the solidity of life, geology, and architecture. It is a most evidently invisible part of our experience. But air is filled with biology and we are its mobile filters, capturing five million fungal spores in our noses and lungs every year,[6] along with one million diatom fragments, other protists whole or broken, bacteria, viruses, and pollen grains.[7] Dandruff and other skin flakes are added to the mixture, and crumbs of scales and feathers, along with fly ash from power plants, soot from automobiles, debris from forest fires, and dust from windblown soil, construction projects, and mining activities. Worship your nose hairs. Pluck or clip them at your peril.

The microscopic things traveling in air are a rarified selection of the microscopic things that live on the ground and in the oceans. There is a passive process of microbial erosion from land and sea as well as a more deliberate component to aerobiology. Airflow strips microorganisms from surfaces and carries them in a predictable fashion over distances that depend mostly upon particle size and wind speed. Some of them remain viable and can resume growth when they settle; others are sterilized by dehydration and ultraviolet light. The deliberate constituents include the spores of fungi, slime molds, and actinobacteria that are adapted for wind dispersal and, in some instances, are catapulted and squirted from their parent colonies.

Dust storms loft billions of tons of soil into the atmosphere every year. Microbes are carried in the dust whether or not they have any adaptations for surviving a ruthlessly dehydrating journey. The driest regions of North Africa are the largest global source of airborne dust, with the expanding deserts of the Sahara and Sahel contributing the greatest quantity of particulates. African microbes in this dust are transported thousands of kilometers and reach the Caribbean and Americas in a few days.[8] The allergenic among these emigrants can irritate the airways of susceptible people at the other end of their journeys, and provocative research links the increased incidence of asthma in Barbados and Trinidad to the growing westward flow of African dust. More and more of us wheeze as the planet desiccates. Dust plumes from Asian storms have been carried even greater distances, with isotopic evidence tracking the transport of topsoil from the cold Taklamakan desert of China across the Pacific *and* the Atlantic to fallout in the French Alps. There are global patterns of transcontinental and transoceanic dust transport in prevailing winds that follow seasonal trends and are exacerbated by drought and deforestation.

Some of the microbes conveyed in plumes of atmospheric dust took their last sips of oxygen thousands of years ago. Every year, ferocious Saharan windstorms carry more than one million tons of fossilized diatoms from the Bodele depression within the bone-dry bed of an ancient lake, or inland sea, called Lake Megachad. Geologists estimate that the winds have scoured a 4 meter depth of Megachad's diatomite deposit in the last millennium, creating gargantuan dust storms and qualifying the Bodele as the world's dustiest place.[9]

Most of this mineral-rich export settles on West Africa and in the Atlantic Ocean, but 20 percent makes it all the way to South America and fertilizes the Amazon basin.[10] Periodically, the clouds of Saharan dust become sufficiently dense to hide the Cape Verde Islands in satellite images. Huge quantities of the dust can settle at the whim of meteorology, caking the islands, and Atlantic shipping, in brown soot. Darwin described this "impalpably fine dust, which was found to have slightly injured the astronomical instruments" of the *Beagle* during her lengthy anchorage in Praia harbor in 1833.[11] Samples were sent to a German expert on protists, who found (according to Darwin), "that this dust consists in great part of infusoria with siliceous shields," namely, diatoms. Most of the diatoms identified in the dust were freshwater species and Darwin concluded that they had originated from the African mainland. Some of the diatoms may have come from the Bodele, but others, deposited on the *Beagle* as unbroken frustules, may have been whipped up from ephemeral lakes elsewhere in Africa. Transatlantic dust storms contain a mix of long-dead and recently deceased algae. The reason that diatoms are so prevalent in these clouds reflects their astonishing abundance in water and the ease with which they become airborne when dried. When a pool loaded with diatoms evaporates, it leaves a fine powder of dry glass particles sized perfectly for air travel.

The smallest of the remnants of African diatoms are capable of making their way into the bronchioles and the alveoli, and might be responsible for some of the lung irritation in the Caribbean populations at the westward end of the plume. It's certain that diatom dust is among the debris circulated in the mucous conveyor belt that cleanses our lungs. Breathing exposes us to a multitude of potential disease-causing microbes, but algae are, with a single exception, absent from this catalog of airborne pathogens.[12]

The harmless nature of photosynthetic protists wasn't appreciated in the nineteenth century, which allowed investigators to make some bizarre claims. James Salisbury was a Civil War physician who became convinced that airborne algae caused malaria.[13] During a severe drought and outbreak of malaria in 1862, Salisbury examined the spit and deeper phlegm of patients suffering from "miasmatic poisoning," and discovered "a great variety of zoosporoid cells, animalcular bodies, diatoms, dismidae, algoid cells and filaments, and fungoid spores."[14] Common to all samples were algal cells "resembling those of the palmellae," which are aggregates of green cells in a jelly matrix.

Searching for the source of these organisms, the doctor positioned glass plates on pegs above stagnant pools in central Ohio and examined the water drops that condensed on the underside of the plates in the morning. He didn't find any of the palmellae in the hanging dew, but reported that they covered the upper surface of the glass. For some time he had no luck in identifying the source of the strange jellied cells. Then, walking across a drying bog in Lancaster, Ohio, he experienced a "peculiar dry feverish sensation." Positioning his plates on the bog he was elated to find the palmellae on the underside of the glass in the morning and discovered the same algal cells, along with "mucidinous fungi," on the crusty soil surface. Salisbury returned to the dried bog in Lancaster repeatedly and experienced the same "feverish feeling" on each occasion. He took an unfortunate colleague on one excursion and was delighted to see him sickened. Algae in phlegm, the same kinds "ague plants" in dew and soil, and feverish sensations when walking over the dried bog: he had found the cause of malaria! (His response to the discovery in 1880 of the genuine blood-borne parasite called *Plasmodium* is not recorded.)[15]

Salisbury established the same link between algae and "miasmatic poisoning" elsewhere in Ohio and in Tennessee. He conjectured that the algae rose into the air after sunset and remained airborne during the night, before descending once more to the soil after sunrise. (This explained why he found the palmellae on top of the plates outside contaminated patches of ground.) With the passage of 150 years, it is impossible to know what he was looking at in the phlegm of his patients, but the location of his "ague grounds" in Lancaster can be determined from his narrative: "between the canal and railroad, and just east of the depot and starch factory." Salisbury's descriptions of feverish sensations sound more like symptoms of asthma, hay fever, or hypersensitivity pneumonitis than malaria. The milling and drying of corn in the starch factory would have posed an occupational hazard for workers, but the three-story building had been repurposed as barracks for soldiers during Salisbury's research.[16] Perhaps he was correct that the symptoms of airway irritation were occasioned by crusts of algae and cyanobacteria left by drying ponds, baked in the sun, and disturbed by farm animals and Union soldiers. The worst symptoms that I experienced visiting Lancaster were caffeine withdrawal caused by the absence of a decent coffee bar. In separate investigations in Lancaster, Salisbury offered unequivocal descriptions of cases of hypersensitivity pneumonitis in farm workers exposed to moldy straw.[17] He made the mistake, however, of linking these cases of allergic illness with an outbreak of measles among the soldiers.

Evidence for the impact of airborne algae on human health is very slim.[18] Diatoms and green algae are certainly in our air supply, but while they may cause respiratory allergies in rare instances, these effects are overshadowed by the potent allergenicity of the far more numerous spores of fungi. Cyanobacteria are a more plausible cause for concern, because some of them produce toxins. The most celebrated case of cyanobacterial damage involves the putative link between a potent neurotoxin, the seeds of cycads, bats, and the prevalence of neurological disease among the bat-eating native population of Guam. The toxic cyanobacterium is a nitrogen-fixing symbiont that resides in the roots of *Cycas micronesica*. Its poison is an amino acid, abbreviated as BMAA, that can induce nerve cell degeneration. Although the bacteria stay in the cycad roots, the toxin makes its way to the seeds.

The problem arises, according to devotees of the bat connection, because the toxin is concentrated in brains of seed-eating flying foxes and people eat the bat brains. The illness in Guam, described as ALS-Parkinsonism dementia complex, is extremely rare in Guam today and its historic decline parallels the extinction of the endemic flying fox. (The last one was shot in 1968.) The chain of neuropathologic causality is fascinating, but isn't strong enough to discourage its critics.[19] A clearer relationship has been drawn between outbreaks of dermatitis and lung irritation resulting from exposure to cyanobacteria discharged from surf during plankton blooms.[20] Though rare, these incidents show the potential harm caused by contact with, rather than the ingestion of, or infection by, noxious bacteria. A handful of studies go further, holding open the possibility that neurological illness *might* result from the inhalation of cyanobacteria.[21]

Any illnesses caused by airborne algae or cyanobacteria are consequences unintended by the microbes. There is nothing in the biology of these non-infectious organisms that favors harm to humans. It is possible that their presence in the atmosphere is similarly unintended. Fungal evolution, as we'll see, has involved the elaboration of an astonishing array of flight mechanisms. None of these are found among the algae. The diatoms driven into the African skies are dead: the majority of the flying frustules are broken fossils, and intact cells blown from drying lakes are dehydrated beyond resuscitation. Additionally, freshwater diatoms are unable to thrive when blown into the sea. But for marine protists able to withstand airborne transmission, escape from the froth of an ocean white cap and reentry after a short flight within water droplets might allow them to access a richer pool of nutrients. There is no requirement for the evolution of processes that augment this passive transportation. I expect that this happens quite often. A few scientists have arrived at a radically different conclusion, positing that microbes form clouds, use them for dispersal over long distances, and may even cause changes in wind speed that get them airborne in the first place. Much of this highly speculative work falls within the pottiest parts of the Gaia Hypothesis.

The argument about microbes controlling weather patterns for their own devices begins with the idea that clouds are products of a biogenic process involving the formation of water droplets and ice crystals around airborne microbes.[22] The bacterium *Pseudomonas syringae* is one of these biological rain makers. *Pseudomonas* infects cereals, peas, beets, and other crop plants. Proteins on its cell surface increase the temperature at which water freezes, inducing ice crystal formation on the leaves of its hosts. The crystals damage the leaves, bathing the bacterium in plant nutrients. *Pseudomonas* is found in the atmosphere, along with other bacteria, and its isolation from hailstones supports the idea that it may be a player in cloud formation. Many other microbes have similar ice-nucleating properties and are found in raindrops and snow flakes.[23] Another biogenic mechanism of cloud formation involves chemical emissions from microbes rather than the physical properties of the cells themselves. Vast blooms of the marine coccolithophorid *Emiliania huxleyi* operate as huge dimethylsulfide (DMS) factories. DMS synthesis is a by-product of the alga's method of maintaining hydration (osmotic regulation), and the compound acts as a potent cloud former over the ocean.[24]

The influence of microbes on weather patterns is a powerful illustration of the importance of the microscopic world in controlling the health of the biosphere, but this does not mean that there is any adaptive significance to these physical processes. In an extraordinary paper published in 1998, however, evolutionary biologist Bill Hamilton and climatologist Tim Lenton proposed that marine algae create their own airflow patterns and clouds to achieve dispersal.[25] They argued that the production of DMS by *Emiliania*, and other marine algae, could cause local changes in wind speed over plankton blooms that would loft the cells from wave caps and carry them over the sea.[26] Putting aside the absence of experiments supporting the operation of the mechanism, its benefit to the alga is problematic. Unlike terrestrial fungi, few algae have invested in the formation of airborne spores. Carriage into the atmosphere is likely to occasion swift annihilation of a cell accustomed to submersion in water. Coccolithophorids might survive brief air exposure, but the evolution of a takeoff and landing mechanism based on the alteration of sea surface and atmospheric chemistry is asking an awful lot from natural selection. The production of a chemical by microbes in the ocean that causes an increase in wind speed that flicks them into the air, then seeds the formation of clouds that carry the cells before dropping them back into the sea in raindrops is a terrifically lengthy extension of phenotype![27]

In the absence any plausible mechanism for deliberate manipulation of the weather by microbes, there are documented instances of the passive airborne transport of living fungal spores across the Atlantic and Indian Ocean. Evidence for these rare dispersal events comes from the sudden development of crop diseases on the opposite side of the ocean from their points of origin.[28] Coffee rust ruined plantations in Sri Lanka in the nineteenth century and followed the crop wherever it was cultivated, throughout India, Java, Sumatra, and the Philippines. A single diseased tree mists the air with billions of the orange spores, and clouds of these infectious particles borne on cyclonic winds spread from the Asian plantations to Africa in the twentieth century.

The Atlantic Ocean was a more formidable barrier. Coffee had been grown in Brazil since the eighteenth century and had never seen the fungus. Quarantines against the importation of coffee plants had protected the Brazilian crop from exposure to the Old World epidemics, but the airborne movement of the pathogen was inexorable. In the late 1960s spores made the crossing from Angola to South America in a few days on easterly trade winds, traveling at a speed of 50–60 kilometers per hour, thousands of meters above the sea. Infected leaves were discovered by a young plant pathologist, Arnoldo Gómez Medeiros, in Bahia in 1970. Medeiros had seen the rust in West Africa during a field trip in 1967, suggesting that the scientist and the spores had passed one another as they flew above the Atlantic.[29]

Sugarcane rust made a comparable equatorial transit above the Atlantic in the 1970s, migrating from Cameroon to the Dominican Republic; novel strains of wheat stem rust moved in an easterly direction from Africa across the Indian Ocean to Australia in the 1960s; and another kind of rust skipped from Australia to New Zealand a decade later. Coffee illustrates the folly of transferring a monoculture from one location to another in the hope of avoiding disease. The atmosphere receives 50 million tons of spores per year from terrestrial colonies of fungi.[30] This weight is carried by an Avogadroian number of spores.[31] Nevertheless, windblown single-step invasions like the transatlantic migrations of rusts are rare events.

Fungal pathogens of animals are less mobile. Dust storms can disseminate *Aspergillus* species that generate lung infections, and valley fever, caused by the fungus *Coccidioides immitis*. Outbreaks of valley fever in the American Southwest are also associated with the dust clouds aroused by earthquakes and construction activities. In addition to the spores of fungi that attack plants and animals, the atmosphere is misted with the spores of saprotrophic and mycorrhizal mushrooms, yeasts related to mushrooms, and the spores of cup fungi and other ascomycetes that feed on just about everything that grows and dies on land.

Traditional methods of identifying and counting these particles rely upon microscopic analysis of slides and air filters. There have been many improvements in the technology of air sampling, but the essence of the research hasn't changed at all since Meier's collaboration with Lindberg. Identification of the spores is a difficult business because some of the most common types are indistinguishable from one another, and cell shape and size can vary a lot within single species. These complications often lead investigators to lump sampled spores into broad categories rather than attempting to discriminate between species. Molecular methods have simplified the process and allow swift analysis of samples at the level of detail—to species, genus, or larger grouping—that suits the researcher. DNA is extracted from particulates captured on air filters, chopped-up into short lengths, amplified, sequenced, and compared with sequences from fungi archived in genetic databases.[32] Classical techniques remain crucial in these guilt-by-association identifications, because the name of the fungus associated with a genetic sequence is always tied to an original determination by someone who looked at the microbe. The microscope will be an indispensable tool for investigating microbial diversity until investigators are content to replace Latin names with numerical codes.

Molecular techniques were adopted in a recent study published by an international group of investigators that took air samples from locations in the middle of continents, coastal ecosystems, and above the ocean.[33] A few of the sea breezes carried no fungi

at all, but spores were everywhere else. Mushroom spores dominated the air from continental locations. The proportion of spores
from ascomycetes increased over coastal habitats and overshadowed the mushroom spores in the air samples collected at
sea. Related studies have examined seasonal changes in aerobiology and track increases in mushroom spores during periods
of maximal fruiting.[34]

The air above rainforests becomes loaded with the spores of
mushrooms engaged in the decomposition of plant debris as well
as mycorrhizal species. In combination with volatile organic
compounds emitted from plants, the spores act as nuclei for the
cloud formation and precipitation that sustain the forest ecosystem. The Amazon basin has been described as a biogeochem-
ical reactor.[35] If the picture of a self-sustaining forest seems
reminiscent of the algae dispersing themselves by controlling the
weather at sea, proceed carefully. The circularity of the mantra
that fungi in rainforests need rainfall so cause rainfall, is broken
by recognizing that fungi capable of initiating clouds grow almost everywhere. Even if the presence of lots of spores above
rainforests influences rainfall, this does not imply that the ice-
nucleating properties of the spore surface evolved to keep the
parent colonies on the ground soaked with water. This makes no
more sense than saying that animals exhale carbon dioxide in
order to support plant productivity.

Fungal spores carry proteins on their surface and allergy to
them afflicts hundreds of millions of people. This is one of the
reasons that I spent a good deal of the last decade examining the
mechanisms that get spores into the air; another is that the feats
of micro-athleticism that separate spores from their parents are
among the most captivating movements in nature. Mushrooms

use a catapult powered by the acceleration of a tiny droplet of fluid over the spore surface to launch spores from their gills; a relative of mushrooms called the artillery fungus employs a snap-buckling device that resembles a miniature toilet plunger to propel a spore-filled capsule into the air, and cup fungi and other ascomycetes use microscopic squirt guns to blast their spores skyward. Most launch devices propel the fungus over a few milli-meters or centimeters, and subsequent dispersal is reliant upon displacement by wind currents.

In recent years we have learned a lot about the way that these devices operate using high-speed video cameras, but earlier re-searchers made huge advances using ingenious experimental ap-proaches. One of my favorites was a 1959 study of squirt-gun ballistics, in which a transparent disc was spun at high speed above a fungus called *Sordaria* that was discharging its spores.[36] When the turntable was stopped, researchers found groups of spores arrayed in short files around the circumference of the underside of the disc. Each file was shot from a single pressurized cell called an ascus that operates as a fluid-filled cannon. The length of these spore deposits was proportional to the time interval for ascus emptying, allowing the investigators to estimate an emptying time of 50 millionths of one second and a launch speed of 11 meters per second, or 40 kilometers per hour. This harmonized with high-speed video observations 50 years later. Another ascomycete fungus is the current record holder for launch speed: spores of *Neurospora* were clocked at more than 100 kilometers an hour using a digital camera running at one million frames per second.

Historical accumulations of fungal spores, as well as spores produced by seedless plants and pollen from the seed plants, offer

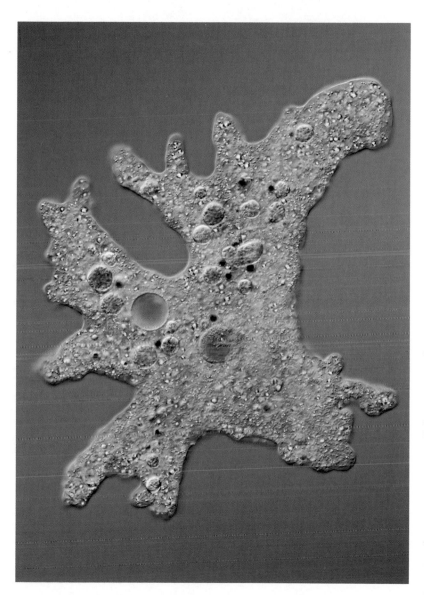

PLATE 1   *Amoeba proteus*, iconic unicellular organism illustrated in evolutionary trees and member of the amoebozoan supergroup.

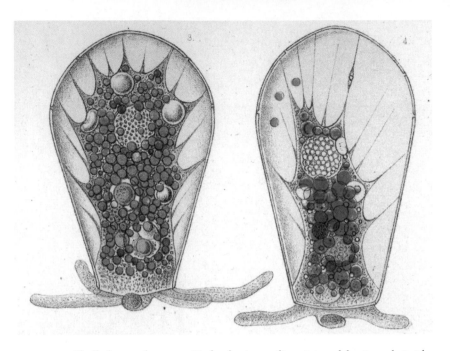

PLATE 2   Shelled amoebozoan, *Hyalosphenia papilio*, pictured by Joseph Leidy. Pseudopodia emerge from the mouth of the shell, or test, and this protective vase is held upright as the cell glides over surfaces. The green globules inside the amoeba are symbiotic algae called zoochlorellae. Leidy was enchanted by this organism: "From its delicacy and transparency, its bright colors and form, as it moves among the leaves of sphagnum, desmids, and diatoms, I have associated it with the idea of a butterfly hovering among flowers."

PLATE 3
Marine coccolithophorid alga armored with spiked scales or coccoliths made from calcium carbonate. Coccolithophorids are members of the hacrobian supergroup.

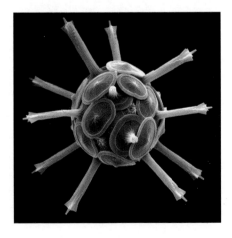

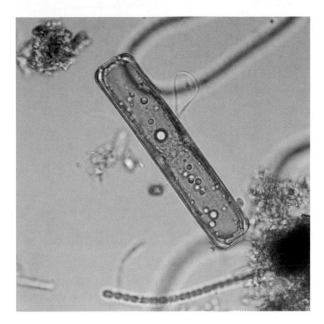

PLATE 4 Freshwater diatom and neighboring filamentous cyanobacteria. The golden-brown pigments of the diatom contrast with the green bacteria. Diatoms are examples of stramenopiles whose membership includes water molds and brown algae.

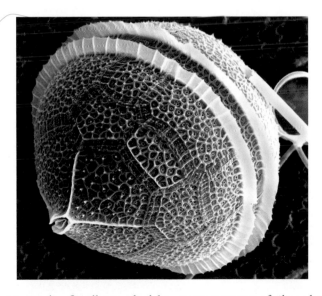

PLATE 5 Marine dinoflagellate with elaborate arrangement of silica plates that sit beneath the cell membrane. The groove around the equator of the cell holds a ribbon-like flagellum and a second flagellum trails behind the swimming cell acting as a rudder. Dinoflagellates are alveolates.

PLATE 6 Symmetrical glass skeletons of planktonic radiolarians. These beautiful frames are embedded in the cytoplasm of living cells and their accumulation in seafloor sediments shows that marine radiolarians evolved more than 500 million years ago. Radiolarians are classified as rhizarians.

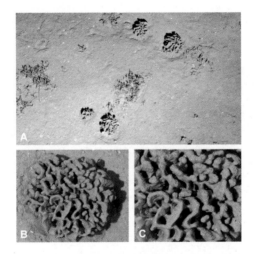

PLATE 7 Giant marine amoebae called xenophyophores photographed at a depth of 4 kilometers in an undersea canyon off the coast of Portugal. Xenophyophores are distant relatives of foraminiferans that belong to the rhizarian supergroup.

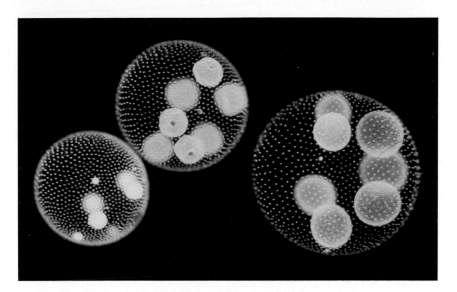

PLATE 8   Mobile colonies of the green algae, *Volvox aureus*, pregnant with the next generation of bright green spheres of this freshwater archaeplastid.

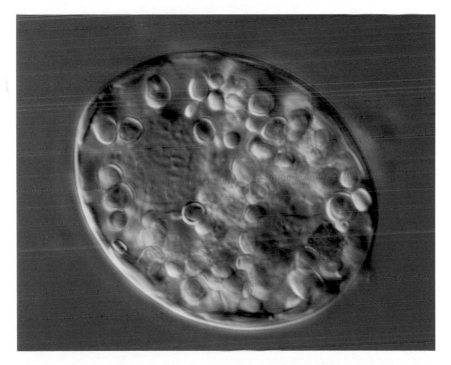

PLATE 9   Freshwater euglenoid alga, *Trachelomonas*, containing bright green chloroplasts and a red eyespot. The cell uses its eyespot to detect optimal locations for light absorption for photosynthesis. Euglenoid or euglenid algae are excavates.

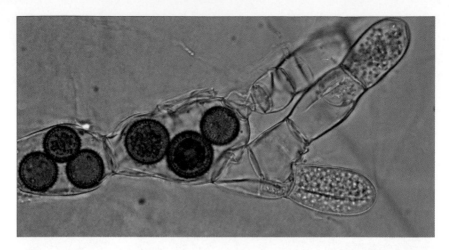

PLATE 10 Spores of the fungus *Rozella allomycis* within infected filaments of a chytrid fungus *Allomyces*. *Rozella* is a member of a phylum of fungi called the Cryptomycota that was defined in 2011 using molecular techniques. The genetic signatures of Cryptomycota are found in rivers and ponds, estuaries, and even in chlorinated drinking water.

PLATE 11 Cluster of choanoflagellates with collars of microvilli. A single flagellum emerges from the center of the collar and its undulations draw food particles to the cell. Choanoflagellates are members of the opisthokont supergroup that unites the animals and fungi. Relative to the great sweep of biological diversity, the choanoflagellates are our close relations.

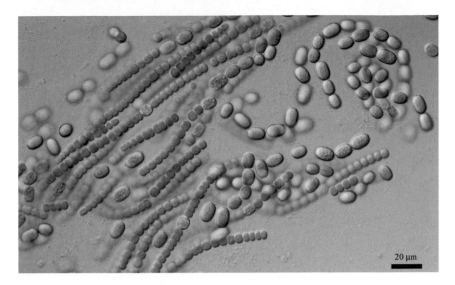

PLATE 12   Filaments of the cyanobacterium *Trichormus variablis*, resembling beaded necklaces. Chlorophyll and other photosynthetic pigments are concentrated in the green cells. The two cells with thick walls within the filaments (left of center) are called heterocysts and function in nitrogen capture, which is essential for the synthesis of proteins and nucleic acids. The chloroplasts of photosynthetic eukaryotes evolved from cyanobacterial cells that were absorbed by the remote ancestors of today's algae more than one billion years ago.

PLATE 13   Model of the influenza virus showing outermost lipid envelope studded with red and yellow glycoproteins, mauve protein capsid, and yellow genome in the center of the particle encoded in RNA strands.

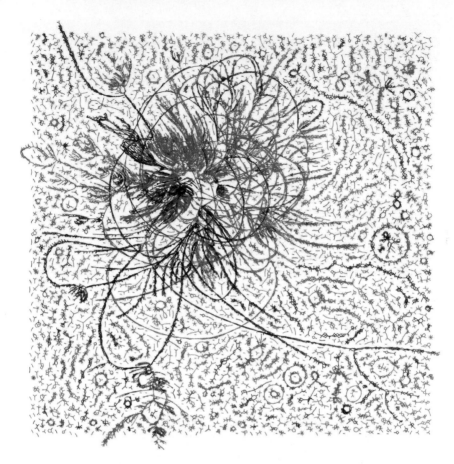

PLATE 14   Metagenomic complexity of surface seawater samples from Puget Sound in Washington State. This colorful graph shows a galaxy of sequence fragments obtained from prokaryotes and viruses and long threads of connected sequences representing the genomes of individual species.

a wealth of information about climate and biology. A fungus called *Sporormiella* that flourishes on animal dung is a major contributor to this palynological record.[37] Changes in the concentrations of *Sporormiella* spores are used as a proxy for the abundance of herbivores. A precipitous decline in these spores in ancient sediments reflects the global mass extinction of megaherbivores including the woolly mammoth, mastodon, and rhinoceros in the late Pleistocene, as well as the seventeenth-century extirpation of species of moa—the giant flightless birds of New Zealand.[38] The profound influence of animal waste on the composition of the airborne microbiome is one of the surprises of this subject of inquiry, and indicates how much we have to learn about the biology of our atmosphere. Even on the subject of the discharge mechanisms of fungi I am discomfited by the necessity of admitting that we don't know how the most allergenic species found in air samples get there. This is a paradox of aerobiology.

Species of *Alternaria* grow on plants living and plants dead. They form elaborate branched stalks on the surface of their decomposing food sources that bear elegant spores shaped like bowling pins elaborated by a glass blower with abstract tendencies. These are conidia, clonal spores whose nuclei are mitotic replicates of the nuclei in the parent colony. Conidia contrast with the spores of mushrooms, and other fungi, that carry genes recombined through a sexual process. The fungus generates clouds of asexual conidia when it begins to exhaust its food source in a particular location. *Alternaria* spores are among the most common spores in nature; most of us inhale them every day and, unfortunately, they are highly allergenic. If you are asthmatic, there is a good chance that your skin will flare red if the spores are pricked into your skin.[39] The stalks bearing *Alternaria* spores are springy and disturbed by the lightest puff of air. The prevailing view of their dispersal process has been that airflow separates the spores from their stalks and they are carried away on the wind. No simpler mechanism is imaginable. The only problem with this solution is that it is not supported by experiments. Airflow over colonies of the fungus grown in the lab fails to release more than a tiny proportion of the spores. Other conidial fungi behave in exactly the same way.[40] Exposure to airflow drives a puff of spores into the air, but most of them stay put on their stalks, buffeted by the breeze but never taking flight.

There are a number of possible solutions to the paradox. Drying of the spore stalk may cause its internal fluid to rupture, forming tiny air bubbles and jolting the spores from their stalks. This cavitation process has been demonstrated in a handful of the conidial fungi, but not in *Alternaria*. Electrostatic charges could be amplified in dry air and lead to the repulsion of spores from leaves. Ideas abound, but compelling demonstrations of behavior are lacking. The significance of this research is heightened by the growing global population of asthmatics—currently estimated at 300 million people[41]—and by an interesting mycological twist to the phenomenon of climate change. Plants fed elevated levels of carbon dioxide in growth chambers grow more luxuriantly, and *Alternaria* colonies infecting their leaves produce more spores.[42] More and more of us wheeze as the planet gets warmer.

Allergenic fungi are common inhabitants of damp buildings. The atmosphere in the indoor environment following natural flood damage, roof leakage, or a plumbing disaster is very different from the outdoor air. The reason for this shift in microbial diversity is that the food sources in a home are different from those outdoors. Drywall, which is a layer of gypsum sandwiched in paper, is a superb food source for a variety of conidial fungi, and becomes blackened and blued with spores when it is soaked. Molecular studies reveal a surprising diversity of fungi in buildings that contradicts the orthodoxy of species richness being concentrated in the tropics.[43] Fungi in the indoor environment exhibit much greater diversity in temperate locations. The latitude of the sampled building was a much better indicator of fungal diversity than its construction materials, suggesting that the diversity of microbes capable of living on wood and paper products, and paint and plastics, is greater in Europe than in Africa. This is interesting in light of a strong correlation between asthma prevalence and distance from the equator, but additional factors including vitamin D deficiency at higher latitudes, and patterns of non-fungal allergens are important too.[44]

Fungi share the indoor and outdoor air with bacteria and viruses propelled inside droplets of mucus at a velocity of hundreds of kilometers per hour from the glistening airways of infected animals.[15] Transmission of most infectious agents is limited to meters rather than kilometers; lengthier journeys in wind are funereal experiences for the majority of microbes whose cells lack the fungal equipment for intercontinental travel. There are a few exceptions, including the reputed spread of cattle viruses in dust storms.[46] Live cells of non-infectious bacteria can be recovered from dust storms, and there is one highly cited and highly suspect report of live bacteria harvested by a Russian rocket at an altitude of 50 kilometers.[47] This is at the top of the stratosphere, well above the protection of the ozone shield. Dust storms whip the microbial world into the stratosphere, but there it dies, its genetic code frizzled by ultraviolet radiation.[48]

Survival below the stratosphere depends upon adaptations to moderately acidic conditions that can develop in clouds, to the frigid temperatures at high altitude, and increasing exposure to ultraviolet radiation. At lower altitudes microbes may grow and reproduce in raindrops, and there is even evidence of bacterial nutrient cycling within clouds.[49] We have seen that broad patterns in fungal diversity can be identified in air samples above forests and other ecosystems, but the opportunities for these organisms to colonize distant locations may be quite limited. Survival in the atmosphere is a tricky business, even for the resistant spores of many fungi, and resettling in a habitat suitable for development is an exceptional event. In this sense, microbial biogeography is largely a matter of environmental and biological conditions on the ground. If airborne dispersal were a dominant factor in determining microbial patterning, we would expect to find the same mix of fungi and bacteria in the soils of Europe or Africa and those of the Americas. This doesn't seem to be true. Wood-rotting fungi whose fruit bodies look the same in different regions harbor surprising genetic endemism.[50] Mycorrhizal fungi have distinctive distributions too, reflecting the importance of host distribution in determining the range of partnering microbes.[51] There is also potential for endemism among prokaryotes, but less is known about their distribution patterns across wide territories.[52] All of these organisms will take advantage of their opportunity of a lifetime when chance lands them in a good spot, but the great mass of microbes circulating high in the atmosphere perish on the wing.

6

Naked Ape

Nor those mysterious parts were then concealed:
Then was not guilty shame: dishonest Shame
Of Nature's works, Honour dishonourable.
Human body, purity or lack thereof
—*Milton, Paradise Lost, Book IV*

We are mobile ecosystems, carrying a galaxy of microbes in the digestive tube connecting mouth to anus, in the moist confines of our genital plumbing and the life-sustaining respiratory system. Our skin is the largest barrier to pathogens, and teems with a fourth protective ecosystem of bacteria and yeasts. Remaining distinctions between *them* (the microbes) versus *us* (cells derived from fertilized eggs) are muddied further by considering the microbial components that constitute our eukaryotic cells as well as the viral ancestry of much of our DNA.[1] Descartes wrote about thinking as proof of the individual's existence, and propagated the ancient belief in our supernatural distinction from the rest of life on earth. Four hundred years later, mainstream philosophy clings to the illusion of human magnificence. Biology has been saying something very different recently; that we embrace the irrational if we imagine that we exist as anything more, *or anything less*, than complicated mixtures of cultivated microorganisms.

And we begin, at the beginning, with your mother's vagina and your slippage though it. Life in the amniotic sac is a coddled preparation for the shock of the first encounter with microbes outside the womb. The fetal immune defenses are tutored for this surprise by a subset of the mother's antibodies that cross the placenta. Gestation shares many of the features of parasitic infection, with the placenta secreting molecules that camouflage the fetal tissues from maternal defenses that might result otherwise in spontaneous abortion. The fetus produces regulatory immune cells that tolerate, rather than attack, the proteins of its developing organs and the larger foreign body of its mother. Once outside, the infant immune system hastens to create a more detailed microbial inventory of its surroundings, with the aid of antibodies delivered in breast milk, categorizing harmful and helpful bacteria.

First contact with swarms of live microbes occurs during birth when the baby is coated with vaginal bacteria. This begins the transformation of the antiseptic fetus into a mammal-microbe symbiosis harboring 100 trillion bacteria. Prominent among the bacteria transferred from vagina to the newborn are species of *Lactobacillus*, and the newborn is dosed with fecal microbes too; bacteria are introduced to the gut by swallowing, to the nose and lungs by breathing, and the immaculate skin is inoculated with bacteria by everyone that holds and hugs and nuzzles and kisses. Every time it attaches to nipples or guzzles from bottles the infant's microbiome expands. Babies born by Caesarian section bypass the vaginal microbiome, and we'll examine the consequences later in this chapter. The increase in the raw number of bacterial cells is accompanied with a steady climb in the diversity of species, along with game changing alterations in the abundance of different groups. Species of *Lactobacillus* and their relatives among the *firmicutes* rule the youngest guts and are fed—*first them, then us*—on breast milk or formula. These cede 30 percent or more of the intestinal banquet to a second phylum, the *bacteroidetes*, as vegetables are introduced to the diet, and these bacteria work with us until we stop feeding them and they stop feeding us.

The first months of life witness great changes in the microbial throng in our feces. The first bowel movements of a baby release the tarry substance called meconium derived from epithelial cells, bile, and amniotic fluid ingested in the uterus. The meconium is low in diversity, rich in firmicutes, and although the breadth of the microbiome increases with each day of life, firmicutes remain in control for the first three months.[2]

The infant ecosystem is structured according to diet, but is agitated, like the occupants of a pond, by all manner of environmental perturbation. The bacteria within the feces of one baby subjected to detailed molecular scrutiny showed spikes in proteobacteria and actinobacteria associated with a fever. Genetic analysis of the eukaryotes in this baby's gut showed that fungi were encouraged by this disturbance, but whether this mycological spike was a consequence of the fever or its cause wasn't clear. Adding another layer of complexity to the picture of gut ecology, it is possible that both bacteria and fungi may have responded to the disturbance caused by an undetected virus. Either way, the exotics vanished in a few days as the infant's fever subsided. Other temporary changes in the baby's microbiome accompanied the end of breast milk and introduction of cow milk at nine months, followed by the prescription of antibiotics for an ear infection.

The most significant and persistent change in the fecal affairs of man accompanies the introduction of cereal and other solid food, when firmicutes cede some of their space to bacteroidetes that possess the enzymatic machinery to process the complex carbohydrates in cereals and vegetables. The microbiome wobbles with each environmental challenge, but reboots with surprising vigor, maintaining the bacteroidetes-rich ecosystem that persists through most adulthoods. There is a smoothness and predictability to these changes that reflects the antiquity of the symbiosis between bacteria and their farm animals (us).

The baby whose microbes were scrutinized in molecular genetic detail was born in the United States, and the ecology of its gut is unlikely to differ very much from the soup inside a Scandinavian or Italian infant. If we look farther afield in terms of dietary practices, and desert the Western reverence for sanitization, the menagerie changes.[3] The ratio of two genera of bacteroidetes—*Prevotella* and *Bacteroides*—reflects the preponderance of plant carbohydrates in the diet. In rural Africa, where maize and cassava dominate the diet, *Prevotella* rules; the balance tips toward *Bacteroides* in a Western diet richer in animal protein and saturated fats. We know that this difference isn't due to a secondary cause because the gut of an American vegetarian looks a lot more like the gut of a Malawian than his or her hamburger-loving neighbor.[4]

These distinctive dietary communities of gut microbes are called *enterotypes*. Adult enterotypes show the same resilience as the infant microbiome, and are unperturbed by anything other than a profound alteration of environmental conditions. Adoption of a low-fat and high-fiber diet is an obvious affront to one's residents if they are accustomed to more fat and less fiber. Even here, however, immediate changes to the microbiome are subtle, and a long-term commitment to the new diet is needed to convert the gut from a *Bacteroides* to a *Prevotella* enterotype.

Radical dietary changes have a more dramatic effect on the gut microbiome, at least in mice converted to a high-fat, high-sugar fast food menu. The guts of these lab animals bloom with firmicute bacteria skilled at processing sugars and the mice become obese.[5] Similar results are observed in humans: firmicutes overwhelm the guts of obese people, and the balance tips back toward the bacteroidetes in response to the adoption of a low calorie diet and the resulting weight loss.[6] The consequences of drastic alterations in diet have been compared with British Petroleum's Deepwater Horizon disaster in the Gulf of Mexico in 2010.[7] The irruption of hydrocarbons into this rich fishery inspired an immediate population explosion among bacteria capable of degrading oil.[8] Immigration isn't necessary in the polluted sea or gut, because the rapid reproduction of preexisting bacteria can bring a minority population to sudden prominence when the environmental conditions change.

Some components of our diet resist processing by our enzymes *and* those of our fecal comrades. There are a lot of calories locked up in the composite of sugary macromolecules in plant cell walls: microfibrils of cellulose constitute the bulk of the structure; these are tethered to one another by hemicelluloses, and the resulting network is embedded in a matrix of pectins. The human genome encodes fewer than 20 enzymes that catalyze the breakdown of this composite and none attack the cellulose polymer. Most of the metabolism of complex carbohydrates is driven by the thousands of carbohydrate-processing enzymes in the gut microbiome.[9] Our bacteria can deal with all manner of chemical configurations among these structures, releasing sugars that power the human ecosystem. Cellulose is the most notable exception. The microbiome generates enzymes that can chop away at one end of the cellulose molecule in the oral cavity, and gut bacteria secrete the complementary enzyme that degrades cellulose from the other end. Separated by the stomach, and lacking other cellulose-degrading enzymes, this combination doesn't allow us to make use of much, if any, of the cellulose, and so most of this fiber is voided undigested. No matter how much roughage we consume, we cannot turn the gut into the kind of hyperefficient fermenter operated by termites and ruminants.

The newness of all of this microbiome research and the implications of our discoveries for revolutionizing medicine are very exciting, and the technology is changing swiftly. The laboratory methods that today's middle-aged microbiologists learned as undergraduates have proven hopeless for illuminating the microbiome. Hints at the richness of the life we carry in our guts came decades ago from the identification of bacteria cultured from fecal samples, but 70 to 80 percent of the gut bacteria perish in Petri-dished captivity and are known only from their genes. Some of the sequences show sufficient similarity to genes from recognized bacteria that it is possible to say, for example, "this comes from a member of the firmicutes related to *Lactobacillus*." Others are so different from known sequences that we can't come close to aligning the organism with an established Latin name. In either case, we refer to the cryptic organisms as operational taxonomic units, or OTUs. There are an awful lot of OTUs in the gut and we may never know them beyond their identification numbers.

Investigation of the human microbiome using shotgun sequencing techniques—the same ones used to investigate seawater, soil, and air—has realized the microbiological equivalent of discovering a coral reef in a rain barrel. First-generation sequencers for reading genes from single organisms, referred to as capillary-based platforms, handle 750-base sequences with very high accuracy. These are perfect for studying the evolutionary relatedness of different species of plants or animals, and the majority of sequencing projects rely on them. They are next to useless, however, when faced with a plethora of unidentified microbes in a dollop of feces. This challenge has been met by the introduction of next-generation sequencing, which can provide millions of shorter reads of 100 to 400 bases at very high speeds and characterize the entire community of bacteria.[10] The rapidity of advances in sequencing technologies is evident from the fact that the cost of sequencing a human-sized genome has dropped from US$95 million in 2001 to less than US$6,000 in 2013.[11] If the same economies applied to the automobile industry, a Rolls Royce Phantom priced at a touch below US$500,000 would cost less than a thanksgiving turkey.

Sequencing studies are augmented by analysis of the transcriptome, proteome, and metabolome. Transcriptomics, or expression profiling, deals with RNA generated from the genome, and proteomics is concerned with profiling proteins translated from the transcriptome. Rather than sequencing genetic information, metabolomic studies identify huge numbers of compounds using mass spectrometry. Mass spectrometers blast molecules into smaller charged fragments, and determine the structure of the original compound by analyzing the debris in an electromagnetic field. This isn't as simple as pushing a button and waiting for the list of chemicals to appear on the computer screen, and there is something of an art in compiling a global profile of compounds in a sample. This research shows that besides their assistance in digesting our carbohydrates, the microbiome engages in an incredible array of metabolic reactions which are critical for our survival.[12] Resident bacteria produce short-chain fatty acids that stimulate water and salt absorption by the gut, inhibit pathogen growth, fuel the epithelial cells of the colon, and provide a spectrum of health benefits; bile acids and choline metabolites from other bacteria regulate the levels of lipids and sugar in the bloodstream, and *Bifidobacterium* (an actinobacterium) produces a string of vitamins to keep us lively.

Despite remarkable improvements in the instrumentation for DNA sequencing and metabolite profiling, the enormous scale of microbiome research makes it very expensive. Analysis of the microbial communities in the sea, the soil, and the air is comparably pricey, but the promise of medical breakthroughs makes the human microbiome the easier sell. Its analysis has become an international research enterprise funded by the NIH Human Microbiome Project in the United States, the European Union's MetaHIT, and the Beijing Genomics Institute.[13] The investment of public and private funds for this work is easily justified. By the time we are 5 years old, most children in the developed world have received multiple courses of antibiotics. Each treatment creates havoc in the microbiome and can affect the natural balance of bacteria for years. The possible consequences of this microbiological misadventure include the gamut of inflammatory illnesses. We need to find out what's happening in the gut.

Antibiotic treatment doesn't change the fundamental enterotype of a person's gut, but researchers have determined that the diversity of bacteria takes a hit from a weeklong antibiotic prescription, and that genes conferring antibiotic resistance are amplified among the surviving bacteria. Neither finding is surprising, but more troubling studies on the aftermath of antibiotic treatment have shown that bacterial diversity remains low and antibiotic resistance persists two years later.[14] Experiments with mice showed that antibiotic treatment altered the levels of almost 90 percent of the metabolites in feces, indicating comprehensive disruption of the performance of the microbome.[15]

The enduring alteration in one's gut ecology in the wake of a single antibiotic treatment is alarming given the growing number of studies linking all manner of human illnesses to alterations in the gut microbiome. *Helicobacter pylori* is the bacterium long vilified for its connections with gastritis (inflammation of the stomach), peptic ulcers, and stomach cancer.[16] Antibiotic treatment to eliminate *Helicobacter* has been heralded as a miracle cure for ulcers, but the bacterium is a normal component of a healthy microbiome too, and its removal by routine antibiotic prescription in childhood is causing concern. By secreting appetite-suppressing hormones *Helicobacter* may aid weight control, and some studies link its inexorable disappearance to the increasing incidence of childhood asthma.[17] *Helicobacter* has partnered with us since our genesis, carried by human populations as they left East Africa 58,000 years ago, becoming increasingly isolated, genetically, just like us, the further we have drifted from home.[18] The forced exodus of this microbe may have serious consequences, and it may be wise to preserve *Helicobacter* strains before they vanish forever.

The list of conditions which may involve our immunological interactions with gut microbes goes beyond inflammatory bowel diseases and includes rheumatoid arthritis, multiple sclerosis, diabetes, atopic dermatitis, and asthma. The first weeks of our lives appear to be the most significant in determining whether or not we develop asthma and other allergic complaints in childhood and beyond. Microbial exposure is a major player here, and this takes us back to birth. Babies born by Caesarian section, rather than the vaginal route, are 20 percent more likely to develop asthma.[19] This may be related to the observation that within the first 24 hours of birth, babies born by C-section are colonized by the kinds of bacteria that grow on the skin surface and miss out on the *Lactobacillus*-dominated inoculum from the vagina.[20] Breastfeeding, as well as more general contact with the mother, supplies *Lactobacillus* to the infant gut, but those first few hours of life when we are presented to microbiology can be decisive. Some research points toward the utility of giving a probiotic to Caesarian babies to make sure they get their *Lactobacillus*. A Finnish study showed that a daily probiotic for babies containing the same kinds of bacteria found in yoghurts conferred some protection against eczema, food allergy, allergic rhinitis, and asthma.[21] How long will it be before obstetricians transfer a dab of vaginal mucus into the mouth of the newborn to be on the safe side?

Experiments on lab animals show compelling links between alterations in the gut microbiome and the development of allergic and autoimmune diseases. Mice given high doses of antibiotics become sensitized to the infamous allergens in peanuts, and show asthma-like symptoms when they are exposed to fungal spores. The implication here is that disturbance of the normal gut microbiome impairs regulation of the inflammatory response of the immune system, resulting in allergic illness.

Another strategy for investigating the relationship between the microbiome and disease is to work with germ-free or gnotobiotic mice raised on sterile food in a sterilized lab facility to keep them free from bacteria. Gnotobiotic mice have to eat more food than normal mice because so much of their diet passes from mouth to anus undigested. The germ-free life confers some benefits, at least for mutant mice that would otherwise succumb to all manner of maladies. For example, mutant strains with variously impaired immune systems become arthritic, develop bone attachment disorders (entheses), and suffer brain inflammation. These ailments disappear, or show greatly reduced severity, if the mice are raised gnotobiotically.[22] Control experiments show that the mice are ravaged by inflammation if the normal microbial circus is encouraged to develop in their formerly pristine guts. The simple conclusion is that the inflammatory diseases associated with mutations in the mouse genome are mediated by microbes in the murine gut. This doesn't mean that we are better off living microbe free; rather, that the complex interactions between the immune system and the gut microbiome—molecular tweets— have far-reaching consequences for animal well-being.

Multiple sclerosis may be another example of the destructive consequences of a good relationship with bacteria gone bad.[23] Multiple sclerosis involves the destruction of the fatty myelin sheaths surroundings nerve cells in the brain and spinal cord. This self-directed attack, or autoimmune response, is performed by T cells that are supposed to be defending the body against viruses. Genetically modified or transgenic mice expressing defective T cells develop a mouse version of MS that begins with paralysis of the tail. The mice avoid this fate if they are treated to a gnotobiotic life, but develop paralysis if they are transferred to cages in which mice with normal guts had been housed and had defecated. Formation of a normal gut microbiota leads to nerve damage. The immune system of the germ-free mice is defective, but, in this case, the faulty defenses do not stimulate disease as long as bacteria are excluded. Another interesting observation is that the severity of the autoimmune response is responsive to the types of bacteria that colonize the gnotobiotic gut. This raises the possibility of treating inflammatory and autoimmune diseases by changing a patient's diet or through probiotics. We may be decades away from unequivocal recommendations on anti-inflammatory diets, but it has been shown that the fresh mix of bacteria encouraged by a switch to a whole-grain diet is associated with reduced levels of the inflammatory cytokine, interleukin-6, in the bloodstream.[24]

Modification of the gut microbiome can be stimulated from either end of the digestive system. The simplest strategy is to guzzle a beverage containing the optimal community members, and trust that the bugs survive the stomach and can begin colonizing the intestine once squirted into the duodenum. This is the unstated principle behind the yoghurts containing live cultures of *Lactobacillus* (firmicutes) and bifidobacteria (actinobacteria) marketed as lifestyle enhancers for women committed to regular bowel movements. The claims made for these probiotic yoghurts are couched in the airiest of terms asterisked to small print confessing the absence of scientific data. On the plus side, yoghurt is not going to cause any harm. Working from the other end of the digestive system, fecal microbiota transplantation, FMT for short, is a more radical approach that is being used already to treat colitis, irritable bowel syndrome, and other inflammatory illnesses.[25] The technique is simple, though not quite as relaxing as drinking yoghurt: a sample of microbiome from a healthy donor is pre-sented to the patient via an enema, through a colonoscope, or via a tube passed through the nostrils into the stomach or duo-denum. The concept is sound. If the patient's microbiome is un-balanced with the growth of pathogenic bacteria in the gut, then reintroduction of the proper microbes may drive the immune system away from inflammatory responses and restore health.

Given that prospects for raising a gnotobiotic civilization seem limited, microbiome therapies may become a standard approach for treating inflammatory and autoimmune diseases. The signa-tures of hundreds of diseases can be read from the imperfections in our personal genomes, but most mutations are not sole arbi-ters of illness. A genetic propensity toward heart disease does not mean that this particular pump will clog. The news from the microbiome is that our fate often lies in the responses of our bac-teria to our immune systems, and vice versa. Therein, fingers crossed, lays a rich harvest of future treatments.

The major studies on the gut microbiome published in journals with the highest profiles, like *Nature* and *Science*, are testaments of scientific prowess. Any scientific paper can be presented in this way, of course, and more than a few of the humblest taxonomic studies of frogs, or whatever, have been fetishized by their au-thors as works of incomparable brilliance. There is something more to the microbiome studies; they are impressive beyond papers on new frogs and mushrooms and wasps. The sheer weight of their information is astonishing. In one *Nature* article, The Human Microbiome Project Consortium identified bacteria from 30 million ribosomal RNA sequences amplified from 5,000 swabs from 242 adults, and read 3 billion DNA bases from each of 700 samples diverted for more detailed analysis.[26]

Like so much of this research, the *Nature* study was conducted by a herd of researchers: the names and affiliations of 248 PhDs and MDs, and MD/PhDs, fill a page of small font type at the end of this publication. The numbers of scientists as well as the bacteria is impressive. Beyond their technical details, the best of the microbiome papers can be appreciated as works of art as well as science. Metagenomic experimentation has demanded the creation of newer and better ways of presenting vast sets of data, and these papers bristle with colored pie charts and bar graphs, molecular trees drawn in circles, and 3D networked analyses that look like star bursts. These data, like the Hubble photographs of deep space, can inspire viewers with the feeblest claims to scientific literacy: look at all of this amazing stuff that lives inside us! There is, as I may have said earlier in this book, more grandeur in this view of life.

Most of the organisms living in the human gut are bacteria, but minority populations of archaea and eukaryotes are phenomenally important residents. Despite the relatively low number of archaea in the gut—100 million archaea versus 40 billion bacteria in every gram of feces[27]—they seem to be crucial to our digestive health. Species that generate methane are the commonest of the gut archaea and the genes of one species, *Methanobrevibacter smithii*, are amplified from fecal samples more than any other. This prokaryote is a strict anaerobe that consumes hydrogen (H_2) as it reduces carbon dioxide to methane (CH_4). Methanogens are found in the guts of 30 to 50 percent of the Western population, but some investigators believe that *Methanobrevibacter* often escapes notice as a sticky film on the wall of the colon. This location doesn't seem very inviting for an anaerobe because it is fed by a richly oxygenated capillary bed. The puzzle may be explained, however, by the assembly of dense populations of other prokaryotes in the intestinal mucosa that furnish the necessary shelter from oxygen. Bacteria in this symbiosis benefit from the consumption of hydrogen by the methanogenic archaea, which creates the perfect chemical conditions to support the fermentation of sugars. And when this complex community is chugging along, releasing ever more calories from our food, we are transformed into bipedal facsimiles of cattle: we get fat and we fart more methane.[28] Could we treat obesity by modulating the composition of the prokaryote community in the gut? Might a probiotic shake remove the methanogens and allow us to keep eating but shrink a bit too? Nobody knows, but you can bet that there is a lot of interest in finding the answer.

In a seemingly contrary observation, researchers found that methane-producing archaea are also abundant in the guts of anorexic patients. The numbers are persuasive: 100 million *Methanobrevibacter* cells per gram of feces inside a lean teen or adult, twice as many in the obese gut, but half a billion per gram from the anorexic.[29] The explanation may be that by starving their gut microbes, anorexics stimulate greater digestive efficiency among their residents. The once marginal archaea help keep the anorexic patient and their microbiome alive. It seems doubtful, to me at least, that discouraging methanogens will do much to treat obesity. Archaeal proliferation probably reflects underlying changes in the gut environment rather than vice versa. We get too fat and the methanogens blossom; we get too thin and the same thing happens.

Other kinds of archaea live similarly mysterious lives in our digestive systems. The weirdest are the salt-loving haloarchaea that live in the mucosal lining of the gut.[30] These were discovered in patients with inflammatory bowel disease, but there is evidence that they live in uninflamed intestines too. Nothing is known about these microbes beyond the fact that the genes of archaea that normally inhabit hypersaline lakes and solar salterns used for crystallizing salt are amplified from the mucosa lining the colon.

Intestinal eukaryotes are as poorly understood as the archaea, but the more we look, the more we find, and representatives of most of the major spokes of the phylogenetic wheel introduced in the first chapter can be found in our feces. Microbial diversity was introduced with an exploration of my garden pond, but we could, just have easily, have found examples of supergroups even closer to home. The exceptions are the archaeplastids and the hacrobians, and this is explained by the reliance on photosynthesis in the majority of these organisms.[31] Like the gut prokaryotes, most of the gut eukaryotes are anaerobes that eschew oxygen, but aerobes are there too, occupying pockets of oxygen (microaerophilic sites) closest to the epithelium and nearby blood vessels.

Blastocystis was thought to be a yeast until scrutiny of its genetics yanked it from the fungi and dropped it into the stramenopiles. It is a strict anaerobe whose cells contain peculiar organelles, related to mitochondria, that power its oxygen-free lifestyle.[32] We acquire cysts of the organism from farm animals and pets, and while fewer than one in ten people in industrialized countries seem to harbor *Blastocystis*, the protist is found in more than 75 percent of people in the developing world. Its activities in the healthy gut aren't clear, and its association with irritable bowel syndrome has led some investigators to regard this relative of diatoms as an opportunistic pathogen.

Many of the gut eukaryotes have a similar status as likely harmless symbionts, termed commensals, which can turn nasty under certain conditions. True pathogens are scattered across the eukaryotes too, including alveolates (*Cryptosporidium*), excavates (*Giardia*), and amoebozoans (*Entamoeba*) that cause diarrhea of varying violence. Fungi are the most diverse eukaryote microbes that can be cultured from our feces, including the ascomycete yeasts *Candida* (also prevalent in the vagina), *Saccharomyces* (baker's yeast and other species), and *Malassezia* (the dandruff fungus and its relatives), as well as filamentous ascomycetes and basidiomycetes. Larger opisthokonts are lacking from the guts of humans in industrialized countries: we tend to be worm-free. And while this seems good from an aesthetic perspective, and removes many causes of debilitating parasitic illness, the eradication of roundworms and tapeworms is another characteristic of contemporary human biology associated with a weakened immune system and the variety of inflammatory illnesses more prevalent among those who enjoy lives of domestic comfort.

It is difficult to determine whether or not these fungi actually grow in the gut, or happen to be passing through, stuck to our foods. The fact that they can be cultured on an agar medium before sequencing is important because it shows that these organisms were alive in the gut; the same isn't true of the mashed tissues of cereal crops and fruits whose genes are plentiful in our feces.[33] What little is known about the ecology of the gut fungi comes from studies on ulcerative colitis. Mice are the experimental models of choice—ours, not theirs of course—and they are fed compounds that inflame and ravage their little guts in a fashion that simulates human colitis.[34] When the mice are mistreated in this fashion they produce antibodies directed against fungi, suggesting that their gut inflammation is associated with the multiplication of fungi. Everything is worse for mutant mice strains that lack normal immune responses because, presumably, they have no means to control their internal populations of fungi. Symptoms of colitis are aggravated in these unfortunate mutants, and the clincher in these experiments is that the effect is reversed by killing the fungi in the guts with an antifungal drug. This finding is reminiscent of the work with gnotobiotic mice that sidesteps the mouse form of multiple sclerosis.

The last component of the gut microbiome is accorded a special name, the virome, and we may know less about this than anything else about gut microbiology. Sequencing of viral genes from feces distinguishes thousands of different kinds of bacteriophage attacking the trillions of bacterial cells.[35] This is comparable to the diversity of viruses in seawater samples, and suggests that viruses may balance the bacterial populations in the gut as they do in the ocean. One in five of the viral DNA sequences was incorporated into the chromosomes of their bacterial hosts, which is a common feature of phage behavior called lysogeny that allows them to coexist with their prey before replication and destruction of the cell. Lysogenic viruses can effect horizontal gene transfer by conveying genes from one bacterium to the next, and can spread novel characteristics among their prey. This mechanism can spread genes for antibiotic resistance among specific types of bacteria in the gut, favoring their persistence when the larger host is prescribed a course of antibiotics to treat an infection of the urinary tract.

When phages are released from their lysogenic phase and burst from their bacterial hosts, the microscopic carnage may be of considerable advantage to us. The human genome doesn't encode enzymes capable of digesting the cell walls of bacteria, which means that sugars, fatty acids, and amino acids incorporated into the gut microbiome aren't available to us until the cell walls of the bacteria are lysed. Products of the decomposition of plant materials by enzymes released from living gut bacteria are available for absorption through the gut epithelium, but the destruction of bacteria by phages probably augments this source of energy.

With growing recognition of the significance of gut microbes in human biology has come the realization that we may learn a great deal about ourselves by looking at the metagenome inside our relatives. In 2005, four years after the sequencing of the human genome, a consortium of researchers announced that they had decoded the chimpanzee genome. The genome of our other closest relative, the bonobo, was completed in 2012.[36]

Given the peculiarity of the modern diet, in comparison with the more limited menu available to wild chimps and bonobos, we might expect to find few similarities between our microbiomes. This isn't the case. Genomic analysis shows that our mix of bacteria is typical of other mammalian omnivores and closest to the ecology of the bonobo gut. Fruits are the dominant ingredient in the bonobo diet and our microbiome hasn't shifted very far from this model.[37] This is surprising given how much animal protein and fat most of us consume, but shows that gut bacteria are always more interested in plant materials. Animal tissues are digested in the stomach long before they reach the bulk of the bacteria that flourish in the colon. Variations in the microbiome between individual humans are far smaller than those between humans and other species. Chimps and the other great apes—gorillas and orangutans—whose diets are richer in foliage than fruits have a microbiome that is intermediate between bonobos and sheep.

Analysis of a wide range of mammals shows that diet, rather than evolutionary kinship, is the determinant of microbiome similarity. The microbiome of gorillas, for example, whose diet is dominated by bamboo, resembles the mix found in horses and rhinoceros, that share a fairly slow movement of food through the gut. There are exceptions to this rule. Pandas, that share the gorillas' predilection for bamboo, don't have a gorillary microbiome. They contain the same kinds of microbes found inside polar bears that eat seals and Arctic explorers. This quirk is explained by the fact that pandas are quite closely related to polar bears and, despite their diet, share the same kind of simple digestive tract designed for swift food processing. Although a panda eats the same type of food as a gorilla, the panda digests the fluid contents of the plant cells rather than fermenting their polysaccharide-rich cell walls.

The gut microbiome of vertebrates is a very unusual ecosystem, unlike anything that exists outside the digestive tract of an animal. The density of the bacteria far exceeds the accumulation of cells in the rest of nature, and the mixture of bacterial types is unique. Metagenomic work on invertebrates reveals a much simpler ecology with communities of gut microorganisms that match various free-living habitats.[38] The complexity of the vertebrate microbiome is shaped by its interactions with the adaptive immune system that is lacking in invertebrates. The intimacy of the immunological relationship with our microbes supports the idea that the adaptive immune system may have evolved to maintain a lifelong partnership with bacteria, rather than working in a purely defensive fashion against potentially harmful members of the external microbiome.

There is more to the human microbiome than the gut, although the majority of our trillions of associates live there. The mucus that coats our mouths is home to thousands of kinds of bacteria. The most prevalent species include harmless strains of *Streptococcus* and *Haemophilus*, along with *Prevotella*, whose dominance is also diagnostic of the human gut in the developing world.[39] As we saw in the digestive tract, disturbance of the microbial ecosystem is associated with disease, and a complex community of bacteria and methane-producing archaea in the mouth is implicated in tooth decay.[40] One hundred or so species of fungi are stuck in the oral mucus too, ranging from *Candida* and other yeasts, to *Cladosporium* whose airborne spores are ubiquitous, and a pathogenic *Cryptococcus* that is benign while it stays in the mouth.[41] The bacteria of the nasal passages are completely different from those of the mouth, with *Staphylococcus* rather than the oral *Streptococcus*, and *Propionibacterium* (implicated in acne) and *Corynebacterium* leading the mix. The lungs support another distinctive microbiome with *Pseudomonas* as the top bacterium. Our largest organ, the skin, is home to other sets of microbes including yeasts that feed on sebum on the scalp and, when overly energetic, cause dandruff.[42] No spot of our outsides or ductwork or plumbing is free from microbes and we are, for the most part, fortunate for it.

Yet research on gnotobiotic mice also shows something surprising about our symbiosis. Germ-free mice are more inquisitive and less anxious than normal mice, as measured by their interest in exploring dark compartments in an experimental box and confidence in negotiating a maze.[43] This riskier behavior is associated with changes in gene expression in key regions of the brain that increase the turnover of dopamine and other neurotransmitters, suggesting that the gut microbiome exerts some regulatory power over the development of the nervous system. Other studies show that the development of a normal microbiome in gnotobiotically raised mice is associated with an increase in serotonin levels. This isn't as surprising as it seems at first because most of the serotonin in the body is stored in epithelial cells lining the gut.[44]

And when we are done, our symbionts eat us, inside out. The body becomes anoxic soon after we stop breathing, creating a once in a lifetime opportunity for the internal oxygen-loathing symbionts to digest the gut wall and migrate into the surrounding tissues. This is evidenced in the bloat stage of decomposition as carbon dioxide, methane, and hydrogen sulfide gases accumulate in the rotting corpse. Fluids in the corpse become frothy with bacteria too, and seep into the surrounding environment as the skin ruptures. Until then: *Caco ergo sum.*[45]

7

Vulcan's Forge and Dante's Hell

As one great Furnace flam'd, yet from those flames
No light, but rather darkness visible
— *Milton, Paradise Lost, Book I*

Patches of plump, barrel-shaped spores dot the leathery leaves of a shrub in the Western Ghats in southern India. All about, the vegetation blazes. Flames are licking through the dry grass. The air around the shrub heats suddenly, boiling the picoliter of water from each spore into the atmosphere. The tiny barrels collapse instantly, rendered drier than the white bones lying on the scorched earth, and bake for hours until the flames die down.

Monsoon rains arrive the next day and the scalded spores rise with the dust from the ashen forest. When the wind drops, some settle on morning-dewed plants that escaped the fire. Within minutes, the spores are rehydrated, have unpacked their food reserves, and commence growth with the protrusion of slender germ tubes.[1] These are Agni's fungi, named after the Hindu God of Fire. Their resilience seems remarkable, and it is, but biology does this sort of thing all the time. Everything is contested every second of its existence. The jellyfish pulsing in shallow water lapping on a sandy beach, and the sunbather reclining with a cocktail: medusa and human prevail alike over tremendous odds. Every organism is an extremophile when the benchmark for comparison is another beast extinguished by the very same conditions that enable the "extremophile" to thrive. The stranded jellyfish perishes on the sand and the tourist drowns in the sea. Everything in biology is somebody's extremophile.

Having said this, it is important to recognize that some organisms are adjusted to environmental conditions that seem, at first sight, to contradict the fundamental statutes of biochemistry. Humans can bear a hot bath at 47 °C (117 °F) for a couple of minutes and survive brief exposure to temperatures above 100 degrees if the air is very dry.[2] *Pyrolobus fumarii*, a species of archaea that lives in the deep-sea chimneys of a mid-Atlantic hydrothermal vent, grows at 113 °C (235 °F) and is chilled below 90 degrees and quits dividing.[3] The function of most proteins is disrupted by heating, but subtle modifications to the structure of proteins in archaea like *Pyrolobus*, along with mechanisms that refold the molecules if they fall apart, preserves their function.[4] *Pyrolobus* was the unchallenged record-holding *hyperthermophile* until 2003 when a related archaea was isolated from a Pacific vent. The usurper was called Strain 121, referring to its facility of growth at 121 °C.[5] This is a game changer for microbiologists, because 121 degrees is the temperature inside a lab autoclave. Autoclaves sterilize surgical instruments, as well as glassware and growth media for microbiological experiments, by replacing the dry air in their steel chambers with pressurized steam. As far as we know, Strain 121 is unique: every other living thing is defeated by autoclaving.

At the other end of the temperature scale, a community of cold-loving, or *psychrophilic* microbes pursues a ferrous lifestyle in rusty red brine flowing from the Taylor Glacier in Antarctica.[6] This outflow, called Blood Falls, pours from a pool of seawater trapped beneath the freshwater glacier and is full of iron and sulfur. The brine is anoxic and refrigerated to minus 5 °C (23 °F). Archaea in this viscous slurry glean energy from traces of organic matter and produce ferrous iron and sulfate. The bottom of Ace Lake in Antarctica is even colder, chilled to a constant 1 to 2 degrees above freezing, and is saturated with methane. The psychrophilic archaea, *Methanogenium frigidum*, lives here.[7] *Methanogenium*, like many psychrophiles, is a methanogen, energized by reacting carbon dioxide with hydrogen to produce methane. Rather than the stiffening of proteins characteristic of thermophiles, the enzymes of psychrophiles are adjusted to maintain the flexibility necessary for catalytic function at low temperature. This adaptation keeps the molecular machinery running, but it does run slowly: *Methanogenium* cells divide once per month, compared with a 20-minute doubling time for the pampered *Escherichia coli* in our guts.

Life persists in much colder places than Ace Lake. Ice cores across Antarctica are full of fungal and bacterial DNA, but there is no sign of archaea.[8] Fungal growth in ice seems very unlikely, and the reason that *Mucor* and *Fusarium* show up here is that their spores are deposited on the surface of the glaciers and preserved at ever-increasing depths as the ice accumulates for hundreds of thousands of years. Bacteria are found in channels that run through the ice, but evidence for their metabolic activity is equivocal. The absence of archaea is puzzling given their evident fondness for the worst imaginable conditions on earth and ability to power ecosystems on the slimmest of organic rations. The record for the coldest temperature on earth is minus 89 °C (–128 °F) and was set in 1983 at a Russian research station above Lake Vostok on the Antarctic Ice Sheet. The best that any microbe can do at this temperature is stay frozen until warmer conditions prevail. The odds of finding life below the ice, in Lake Vostok's liquid water, are a little better. At a depth of 4 km below the station, the deepest layer of 400,000-year-old ice melts into the lake whose pristine water is warmed to a balmy –3 °C (27 °F).

Lake Vostok has been isolated from the rest of the biosphere by its frozen lid for 15 million years. The stability of the subglacial climate, as well as the absence of immigrants and impossibility of emigration, will have buffered the ecosystem—if there is one—from the changes that affect life above the ice. News reports have referred to Vostok's "Lost World," and suggest that a garden of antiquated microbes awaits discovery. It is sensible, however, to consider that the passage of the last 15 million years has consumed only 0.4 percent of the history of biology. In a two-hour movie summarizing 3.5 billion years of life on earth, sunbathing organisms in the waters of Lake Vostok would be covered with ice in the final 30 seconds. Vostok doesn't offer much of an opportunity for time travel. Little will have changed, speaking microbially, in such a brief period of time. Consider the impossibility of reconstructing *Citizen Kane* from the final scene of Rosebud's destruction and you have an idea of the difficulties facing experts in microbial evolution trying to figure out how the great wheel of life was constructed.

The more interesting question is whether the lake harbors any life at all. Temperature is the least of the concerns for prokaryotes in this environment. Besides the lack of solar power, Lake Vostok is poisoned with supersaturated oxygen levels, and pressurized to 350 atmospheres.[9] Against opposition from scientists concerned about contaminating the lake, a Russian drilling team reported that they had penetrated the overlying ice sheet in 2012.[10] What will they find? Microbes have been harvested from younger subglacial lakes, but Vostok is a different beast, the queen of antediluvian iciness.[11] If anything can thrive in this Dantean Hell—for the poet dispensed with flames and imprisoned Satan in ice—prospects for biology in the subsurface ocean on Jupiter's moon Europa seem a little brighter.

Cold *marine* environments are also gardened by microbes. As seawater freezes, its salts become concentrated in the shrinking liquid phase and lower the freezing point of the remaining fluid. Arctic and Antarctic sea ice is shot through with tiny veins filled with brine and teeming with organisms. There is some evidence that bacteria in these networks remain active even when the temperature drops toward minus 20 °C (−4 °F).

Thousands of meters below the ice, the cold bed of the deep sea catches a dribble of organic matter from the water column—the last flakes of marine snow—along with the occasional whale carcass. Despite the apparent slimness of the pickings, the sediment beneath this abyssal plain is a prokaryotic paradise that some microbiologists contend houses one third of all of life on earth.[12] Uncontaminated retrieval of living organisms from hyperpressurized sediment several kilometers below sea level isn't a cakewalk, but it has been done.

Oil exploration companies have provided vital support for numerous research ventures, and drilling projects for purely scientific purposes have been in operation since the 1960s. An international endeavor called the Integrated Ocean Drilling Program (IODP) is supported by 25 countries that have contributed to the continuous exploration of the subseafloor environment for the last decade.[13] Live prokaryotes retrieved from sediment depths of up to 400 meters include sulfate-reducing bacteria, methane-generating archaea, and species of more familiar bacterial genera including *Bacillus* and *Rhizobium*.[14] Cultures of the fungus *Penicillium* have also been grown from sediment sampled 127 meters below the floor of the Pacific Ocean.[15] This is a mystifying discovery, because we have no idea what a filamentous fungus is doing down there with no oxygen and so little to eat.

Most of the eukaryotic DNA found alongside the bacteria and archaea is thought to be fossilized material from cells that died and descended from the water column. Rates of decomposition in the seafloor sediments are exceedingly slow, and the original owners encoded by these nucleic acids were buried as long as 10 million years ago.[16] Generation times of the prokaryotes living in this low nutrient "Inner Space," or "Deep Biosphere," range from years to thousands of years. This makes the *Methanogenium* from the freshwater Ace Lake seem positively hurried! The terms "microbial couch potato" and "slo-mo life" have been used to describe these slothful organisms.[17] Intact cells of archaea have been recovered from a maximum sediment depth of 1.6 kilometers, where the absence of mixing suggests that these populations are more than 100 million years old—buried since the Cretaceous.[18] Below this depth, sediment temperature rises above 100 degrees, so everything is a hyperthermophile. Theoretically, the inhabitable sediment could extend to a depth of 3 or 4 kilometers before heat radiating from the planet's core devours the last slivers of biology.[19] Future drilling projects will examine this hypothesis.

Although the seafloor sediment is dark, its leisurely residents are powered indirectly by sunlight absorbed by the photosynthetic diatoms, coccolithophorids, cyanobacteria, and other chlorophylled plankton whose remains sink from the seawater above. Where the ocean floor is perforated by hydrothermal vents (hot vents), cold seeps (cold vents), and other kinds of crustal wound, the energetics are very different, and highly productive ecosystems are supported by chemotrophic rather than phototrophic metabolism.[20]

Clams, crabs, shrimp, eels—most dramatically, the slippery congregation called "Eel City" around an underwater volcano in the Samoan Islands—and giant tubeworms, get most of the attention in television documentaries on vents. The phallic forests of 2-meter-plus long tubeworms, *Riftia pachytila*, are, however, mere decorations for vents that spew super-heated geofuels and feed high densities of prokaryotes including thermophilic archaea. Most of the biomass at the vent exists in single-celled organisms, and even the worms are symbioses whose weight is equal parts worm and chemotrophic bacteria.[21]

Geofuel composition varies according to vent location, and the resident microbes live by stripping electrons from dissolved hydrogen, methane, hydrogen sulfide, ammonia, ferrous iron, and manganese as the hot fluxes mix with the cold water. The tubeworm bacteria, for example, synthesize sugars from carbon dioxide using the reducing power of hydrogen sulfide that boils from vents called *black smokers*. Lower-temperature vents called *white smokers* spew an antacid froth of calcium, barium, and silicon into the seawater. According to a provocative theory, life may have originated in specific manifestations of this kind of vent when the ocean was more acidic than today. Flowing through a honeycomb of pores throughout the mineralized chimneys, the alkaline discharge from the vent would have created a natural proton gradient when it met the cold seawater—a geological facsimile of the charge separation across cell membranes that is the essence of every living thing. Operating in this fashion, the vents created an abiotic template for the first cells that might have assembled themselves upon this preexisting chemistry.[22] And here, 3.5 billion or so years after life percolated in the vent chimneys, I sit typing this page and you'll be dozing in your armchair.

Microbes on the planet's surface also cope with an amazing range of environmental challenges. Geothermal springs, steaming fumaroles, jetting geysers, and bubbling mud pots provide microbes with some of the same chemicals expelled by hydrothermal vents. Yellowstone National Park was described by an early visitor as "the place where Hell bubbled up," but it is a place of incredible microbial diversity and has been a center for extremophile research for decades.

There are far tougher places to scratch a living than Yellowstone's geysers, including Pitch Lake in Trinidad, which offers the worst imaginable conditions for life on land. Oil seeping through a deep fault in the earth's crust spreads over an area of 46 hectares, forming a desert of warm liquid asphalt. As the oil migrates toward the surface, the lighter components evaporate, creating a natural form of the tar used for road surfacing. Big bubbles of methane bloat and collapse at the viscous surface of the most active sites, where the temperature rises to 56 °C (133 °F). In other areas the asphalt cools and solidifies.

Trinidadian tar is full of life. Each gram of gunge contains one to ten million bacteria and archaea, which is comparable to the density of prokaryotes in forest soil.[23] This is mystifying because there is very little water in liquid asphalt. The most likely explanation for the vitality of the asphalt microbes is that they reside in brine-filled microhabitats within the tar, rather like the organisms that grow in sea ice. Tar microbes include archaea that oxidize the methane, and metal-breathing archaea called thermoplasmatales that use ferric iron and manganese, rather than oxygen, as their terminal electron acceptors. Bacteria are more plentiful than archaea, and they drive an active sulfur cycle with separate species oxidizing and reducing sulfur. Other bacteria feed directly upon the hydrocarbons and include strains capable of degrading the heaviest oil. Prokaryotes that generate methane are absent from the asphalt, but plenty of them live nearby in a mud volcano called Devil's Woodyard.

Pitch Lake is a site of unique geological and biological strangeness; the only places that offer comparable riches are Lake Guanoco in Venezuela and a trio of tar pits in Southern California, including the famous Rancho La Brea in Los Angeles. Just as exobiologists are excited by the possibility of life in the nippy water of Lake Vostok as a reflection of life's possibilities on Jupiter's moon Europa, the work in Pitch Lake encourages dreams of flourishing ecosystems supported by the methane rain on Saturn's moon Titan. The problems for biology on Titan, however, include a surface temperature of minus 179 °C (−260 °F), which would stop any conceivable enzyme from encouraging a biological chemistry.

Microbial communities within oil reservoirs are a source of great concern to the petroleum industry because they remove the saturated and aromatic hydrocarbons that define light crude oil, and produce more viscous fluids that are more difficult to extract and less profitable at the refinery. This is ironic, in the sense that oil and gas are derived, in the first place, from the slow decay of accumulations of planktonic microorganisms beneath the ocean and freshwater lakes. Bacteria and archaea generate methane, carbon dioxide, and water as they decompose the plankton and, in combination with non-biological processes, form a waxy material called *kerogen* that is transformed into crude oil and gas as the hydrocarbon purée is buried deeper and deeper.[24] At depths of a few kilometers, the deposits are heated beyond the threshold for microbial activity, and remain bottled for millions of years until geology or humans bring them back to the surface, where the microbes are poised for another series of chemical transformations.

Other forms of extreme surface environments include highly acidic refuse piles associated with coal mining and alkaline soda lakes. The greatest *acidophile* lives in a geothermal spring in Japan. This is *Picrophilus*, an archaea that can grow at pH 0.06—equivalent to battery acid.[25] The survival of *Picrophilus* hinges on the impermeability of its cell membrane and the bailing action of highly efficient molecular pumps that expel protons from the cytoplasm. Anything capable of growth under these caustic conditions reaps the physiological benefit of access to a huge acidity gradient that powers food uptake even if nutrients are highly diluted in the surrounding water. Among the eukaryotes, a single photosynthetic alga (a red archaeplastid, *Cyanidium caldarium*) and a few fungi grow at very low pH, but acidiophily isn't a very popular way of life.

At the other end of the pH scale, *alkaliphilic* archaea can grow at a maximum pH of 11, matching household ammonia and bleach. These conditions are found in soda lakes that have a very high salt content. Mono Lake in California is the best-known soda lake in the United States. Organisms in soda lakes face the opposite problem to acidophiles: in a proton scarce habitat, they have a tough time importing enough hydrogen ions to create the slightly acidic conditions necessary for cytoplasmic function. This is solved by a series of physiological contrivances that produce metabolic acids, structural modifications that trap protons at the cell surface, and the deployment of specialized transport proteins that maximize proton import through the cell membrane. Living in a soda lake also presents the challenge of dehydration when bathing in brine. This *halotolerance* is achieved by accumulating salts and sugar alcohols within the cytoplasm to buffer the cell against dehydration through osmosis, and by synthesizing oddly structured proteins that remain soluble inside the salt-rich cell.[26]

Tolerance to extremely high salt concentrations is associated with some curious alterations in cell shape. A quick thought experiment shows that most prokaryotes can be sliced in a way that exposes a circular or oval cross-section: slivers of cocci are circular or oval whichever way they are cut, and rod-shaped bacteria have a circular midsection. These smooth shapes reflect the simple geometries of balloons inflated by pressure, because most bacteria live in quite dilute fluid environments where osmosis drives water into the cell. For archaea that inhabit salt lakes and solar salterns—pools where evaporation is used to crystallize salt— things are very different. Even when they buffer themselves against complete dehydration by accumulating high concentrations of salts and sugars, there is only a minimal gradient to encourage water influx. *Haloarchaea* survive in very salty places and they have very strange shapes: these are archaea with square and triangular cells whose drips of cytoplasm are squished into tiny flat envelopes.[27]

There are a couple of possible reasons for the development of eccentric shapes among these microbes. The first hypothesis is that the loss of internal pressure has allowed these cells to escape the usual physical constraints upon morphology and elaborate remarkably angular shapes. A more satisfying explanation is that the shape of a flattened square more than doubles the surface area to volume ratio, which pays dividends by affording the cell membrane much greater access to the cytoplasm and control of its water content.[28] Microbes bathing in solar salterns absorb an awful lot of ultraviolet radiation and the resilience of fungi in this environment is due, in part, to the melanin pigments that polymerize in their cell walls and act as a natural sunscreen. Black yeasts are the most common fungi growing in these saline habitats.[29] One of these, *Hortaea werneckii*, displays astonishing physiological virtuosity by growing with or without salt and maintaining growth at up to 4.5 molar sodium chloride (seawater is 0.5 M NaCl). Whether the pigmented cell wall of this yeast operates as a salinity barrier, or sunscreen, or performs both functions, isn't known.

Melanin is also characteristic of fungi that tolerate ionizing radiation from radioactive elements. Although ionizing radiation is a significant stimulus for evolution, exposure to a lot of it is a damaging thing for the individual. Sensitivity varies a great deal: humans are killed with a single dose of 5 Gy, or grays, equal to 5 joules of energy per kilogram of body mass. Firefighters at Chernobyl Atomic Energy Station in Ukraine who received doses of 6–16 Gy developed radiation sickness within minutes of exposure and all but one of them died.[30] The remains of the fourth reactor at Chernobyl and more than 200 tons of highly radioactive waste have been surrounded by a concrete sarcophagus since the disaster in 1986. Inside this tomb, melanized fungi coat the walls and ceilings and proliferate along cable passages, where they receive hundreds or thousands of grays per year.[31] Immediately after the explosion, melanized fungi replaced non-pigmented species in forest soils surrounding the reactor, but non-pigmented forms made a comeback in the following decade.[32] Far from tolerating the high doses of radiation within the sarcophagus and in the fallout-contaminated woods, pigmented fungi seem to thrive in these radioactive habitats. Some of the fungi seek hot particles of graphite discharged from the reactor and grow toward radioactive phosphorus and cadmium in the lab.[33] They are positively *radiotropic*.

Other experiments show that heavily pigmented fungi that cause human infections show faster growth rates when exposed to radionuclides. The growth stimulation may be due to warming of the cell caused by absorption of the radiation by the melanin. A more provocative suggestion is that these fungi engage in an enigmatic form of energy harvesting that resembles photosynthesis, where melanin replaces chlorophyll, absorbing radiation rather than sunlight.[34]

The bacterium *Deinococcus radiourans* withstands 5,000 Gy, and more than one third of its cells survive exposure to 15,000 Gy. It is honored with the title, "Conan the Bacterium."[35] DNA is DNA, and Conan's is just as vulnerable to being fried by gamma rays as our genetic code. This is a fact of chemistry. The difference lies in the organization of the bacterium's genetic information and its remarkable repair mechanisms. Each of the berries (*Deinococcus* means strange berry) contains four or more copies of its single circular genome. Because radiation damage doesn't occur at the same places on every copy of the chromosome, it is likely that a coding region molested on one copy of the genome lies unsullied on another chromosome. Intact copies of genes serve as templates for the restoration of damaged sequences. The cell repairs its genome using a protein identified as RecA that identifies and mends breakages across both strands of the DNA double helix.

Every organism has a version of this vital protein. Ours is called RAD51, and if its function is impaired we suffer genomic instability and increased susceptibility to cancer. The RecA protein of *Deinococcus* works in a peculiar reversal of the usual series of reactions, and dispenses with some of the finer points of the molecule's functions in favor of grabbing and mending the broken DNA as swiftly as possible when it is under radiation attack.[36] In addition to its molecular defenses against radiation damage, *Deinococcus* has an unusually thick wall which may hold the cell together. The strength of its fortifications is illustrated by experiments in which a glass beaker containing a culture of *Deinococcus* turned brown and brittle when it was placed next to a block of radioactive cobalt, while the culture endured the gamma rays, repaired its mutations, and emerged unblemished.

Natural exposure to radioactivity is far weaker than the punishment hurled at *Deinococcus* in the lab, which raises the question of the evolutionary significance of its astonishing capabilities. Why invest in a thick cell wall, multiple copies of its genome, and special repair mechanisms that it never uses? The bacterium seems, without deeper inquiry, to be the prokaryotic likeness of a ballerina encumbered with a snorkel and flippers: what on *earth* is the cell preparing for?

The answer lies in the similarity between the molecular havoc created by radiation and by severe dehydration. When a cell is dried to a crisp, its contents are compressed; when water returns, the macromolecules, including DNA, are wrenched apart as the cytoplasm is transformed from a salty smudge to moisturized gel. This disturbance subjects the DNA molecule to considerable shear forces that can break one strand, or both strands, of the double helix. So, like ionizing radiation, water stress damages DNA. Much the same can be said for having a strong cell wall. The other forms of environmental stress discussed in this chapter challenge cells in much the same way: resistance to forest fires is an issue of dehydration tolerance; survival in brine channels in ice requires substantial resistance to osmotic drying; and the greatest challenge of the asphalt desert is its lack of water. Providing a cell can hold on to some water, its metabolism can proceed as long as its proteins are not frozen into submission or denatured by heat.

And this seems to be the reason that *Deinococcus* is resistant to radiation: its defenses evolved as a response to desiccation. Ecological evidence for this idea comes from the observation that *Deinococcus*, and bacteria with comparable *radioresistance*, tend to live in very dry soils, including the driest of the dry, the McMurdo Dry Valleys in Antarctica. These are cold deserts whose permafrosted terrain is parched by fierce winds that cause snow to sublimate into the atmosphere before it touches the ground. Glacial melt during summer months liberates a little water, but this is the coldest, driest place on earth.

Nevertheless, rock surfaces are colonized by photosynthetic cyanobacteria, and a surprising diversity of filamentous fungi and green algae live inside the porous structure of the sandstone (these are called *endoliths*) as well as within cracks that run through the rocks (*chasmoliths*).[37] The cyanobacterium *Chroococcidiopsis* is the most common endolithic microbe, and *Deinococcus* inhabits the bone-dry mineral soil. Both prokaryotes are resistant to desiccation *and* radiation. Ultraviolet light is another source of environmental stress for the microbes of the Antarctic and is a significant cause of DNA damage. Mechanisms used to repair DNA after irradiation with gamma rays and severe dehydration also equip the cells to cope with the intense UV exposure on the surfaces of rocks and soil in the Dry Valleys. Exobiologists draw faint encouragement from these Antarctic ecosystems, looking to the surface of Mars and wondering if anything lived there during its glaciated past. Among the multitude of environmental challenges on our neighbor, cosmic radiation excludes any Martians from the surface.[38] The absence of volcanic activity is a problem for subterranean ecosystems too, and this makes the cryovolcanoes at the south pole of Saturn's moon, Enceladus, a more appealing home for extraterrestrials.[39]

Features of cold and hot deserts, hydrothermal vents, sea ice, and other places that seem marginally encouraging for biology are replicated in our homes, and some unusual microbes share our living space.[40] Even if we have never slept with them, we are acquainted with the horrors of bed bugs, fleas, and lice, and all of us share our bedding with an extraordinary range of bacteria and fungi adapted to the exigencies of overnight warming and daytime cooling. They are warmed by our bodies and fed by skin flakes, hair strands, earwax, pillow drool, and other emissions.

Bathrooms are veritable breeding grounds for thermotolerant microbes that adjust themselves to the daily swings in temperature around a shower head, desiccation-resistant xerophiles that exploit the mist as it condenses on a shower curtain, and alkaliphiles that endure a slosh of bleach into the toilet bowl. Bathroom nutrients come from soap and shampoo residues as well as the debris from our depilations. Even after a thorough cleaning with antimicrobial sprays and scouring agents, the colonies resume growth as soon as the next visitor reignites the microbiological Eden with her ablutions.

The laundry room fertilized with alkaline washing liquids is the domestic approximation of the deep seafloor microbiome, and the kitchen is another study in microbial diversity—from the sheen of bacteria covering the counters, to the fungi growing on food scraps and the psychrophiles reproducing in the refrigerator. Dishwashers harbor unexpected microbial diversity, and some of it is unpleasant. Two species of the disease-causing fungus *Exophiala* are found in more than one third of dishwashers.[41] These ascomycetes grow as filamentous hyphae or as yeasts, colonize the lungs of cystic fibrosis patients, and, on an occasional basis, cause fatal brain infections. The good news is that these fungi live in many other locations, and that the dishwasher is an unlikely source for an infection. Most of the challenging mycoses occur when our immune systems are impaired, but, once in a while, an otherwise healthy individual will succumb to a dreadful infection and will never learn why. The microbes in our homes are no more dangerous than the microbes outdoors. Every day of our lives, everywhere we go, we are filled with and immersed in a soup of invisible microbes. A hot shower and dental scrub with an electric toothbrush change none of these fundamentals.

Extremophiles offer a wealth of potential solutions to vexing problems in biotechnology, which drives a lot of the research on these microbes. Enzymes from thermophilic archaea, and bacteria that catalyze reactions at high temperatures, can replace costly production processes that rely upon conventional chemistry. Examples of these applications include the synthesis of high-purity pharmaceutical agents using enzymes from archaea to prune problematic chemical groups from the precursors of chemotherapeutic agents.[42] Molecular engineers dispense with the archaea themselves and use dense cultures of E. *coli* to express the foreign proteins at sufficient levels for further experimentation.

Other compounds from archaea show promise as heat-stable antibiotics and anti-tumor drugs. Commercialization of these compounds has been very slow, but enzymes from thermophiles have already enjoyed universal application in molecular biological research. DNA polymerase from the bacterium *Thermus aquaticus*, called *Taq* polymerase, was the original enzyme used for the polymerase chain reaction (PCR) by Kary Mullis in the 1980s.[43] The bacterium was isolated in Lower Geyser Basin in Yellowstone in the 1960s, and the ability of its enzyme to copy DNA at high temperatures revolutionized molecular genetic research, changed the way we study biology, provided forensic investigations with a stronger scientific foundation, and affected medicine forever. DNA polymerases from other thermophiles have greatly improved the fidelity of PCR products because they incorporate a proofreading capability in their structure.

Most of the organisms regarded as extremophiles are prokaryotes. Agni's fungi described at this beginning of this chapter survive almost instantaneous dehydration and equally swift rehydration. No fungus competes, however, with the thermophilic performance of Strain 121 and other archaea. Eukaryotes can survive heating, but they don't seem to grow at temperatures above 60 °C (140 °F).[44] The cellular construction and large genomes of eukaryotes present an inherent fragility compared with the smaller and simpler everything of bacteria and archaea, and this excludes them from the places occupied by the most extreme of the extremophiles.

Looking at the wheel of eukaryote diversity, it is clear that some groups are ill fitted, or have not fitted themselves, to a tremendous range of environmental circumstances. In the reverse of the order introduced in the first chapter, we begin with the opisthokonts. Fungi are the paradigm of eukaryote extremophilia, and we have seen that animals, supported by symbiotic prokaryotes, thrive in hydrothermal vent communities. This supergroup meets the criteria for extremophilia.

Excavates are poorly represented in the classic examples of extreme environments, but the disease-causing trypanosomes and *Giardia* in this supergroup qualify as extremophiles if we stretch the definition to include parasites that overcome enormous obstacles in the form of the immune defenses of their hosts.

Fed by photosynthesis, the archaeplastids are restricted to the land and to the surface of aquatic ecosystems. Red algae are the hardiest of the supergroup, with one calcified species living at a record depth of 268 meters in the Bahamas, where light penetration is a few millionths of the level at the surface.[45] To researchers in a submersible who switch off the lamps, the seamount habitat of this alga is dark as a coal mine. The alga absorbs photons in this unlikely garden using phycoerythrin pigments that harvest energy from the weakest light whose blue wavelengths make it this deep.

The extension of the extremophile qualification to parasites embraces some of the SARs (stramenopiles, alveolates, and rhizarians), and cryptomonad algae living in snow allow the hacrobians to claim psychrophilia. This leaves the amoebozoans, and, again, there are plenty of parasites that make the grade. This glance at the environmental challenges mastered by eukaryotes shows the difficulty in separating the *real* extremophiles from *virtual* ones: everybody is somebody's extremophile.

An obscure group of rhizarians deserves a special mention in the parade of eukaryotic extremophiles. *Xenophyophores* are giant marine amoebae that live in the deepest recesses of the ocean, studding seamounts and the seafloor with stalked fans as big as dinner plates, lumps, sponges, and sieves embedded in the ooze.[46] These structures are organized by enormous single cells that trap, organize, and glue particles, including their own waste materials, onto their surface. The stalked forms were discovered by Ernst Haeckel who analyzed protists and invertebrates collected during the circumnavigatory voyage of the *HMS Challenger* in the 1870s.[47]

Haeckel thought that these strange organisms were sponges, missing their affiliation with the planktonic foraminiferans (rhizarian supergroup) whose beautiful cells he showcased in his sensational illustrations.[48] Xenophyophores have been videoed at a depth of 10.6 kilometers in the Marianas Trench by oceanographers at the Scripps Institution using an untethered "Dropcam." These species look like elaborate drain covers, plates perforated with symmetrical arrangements of holes connecting to the larger amoeba beneath the sediment. They are thought to filter feed through their sieves, and other kinds absorb food particles like conventional amoebae using pseudopodia. The pressure in the trench is one thousand times higher than the pressure at sea level, which means that xenophyophores, like all of the deep-sea microbes, are *barophiles* or *piezophiles*. The only civilian submariner to reach the depths of the Marianas is filmmaker James Cameron, who completed a record-breaking solo dive in 2012 in a vertical torpedo called *Deepsea Challenger*.

Xenophyophores rank among the largest cells, if we include single bags of cytoplasm containing multiple nuclei. Multinucleate cells are known as *coenocytes*. Plasmodia of myxomycete slime molds (amoebozoan supergroup) that can ooze over a tree stump, and 1,000-hectare colonies of fungi, are examples of other kinds of coenocyte that can get very big. Siphonaceous green algae (archaeplastids) can get big too. The single-celled alga *Valonia* is a bright green sphere with a maximum diameter of 5 centimeters and has the common name of "Sailors' eyeballs." The cytoplasm of this alga is limited to a thin layer beneath the cell wall, and a huge vacuole occupies most of its volume. Meter-long single-celled rhizomes of *Caulerpa* species form branches that range from clusters of bubbles, trumpet-shaped stubs, and "leaves" that resemble tiny sago palms. Single cells of *Halimeda* are bigger still, forming meadows of calcified fronds in the Great Barrier Reef. Coenocytic amoebae, plasmodia, fungal colonies, and giant algae are the largest of the unicellular organisms discussed in this book.[49] They belong to groups of microorganisms but are disqualified as microbes, which raises the question of the meaning, or worthlessness, of the name, "protist." We'll address this in the final chapter. While we're on the subject of size, it is very rare for a prokaryote to be visible to the unaided eye. The sulfur bacterium, *Thiomargarita namibiensis*, is a rare instance of a prokaryotic giant, whose cells have a diameter of 0.75 millimeters.[50]

Intimate physical mutualisms permit microorganisms to work in conditions that neither partner can abide alone. The lichen symbiosis between fungus and alga, or fungus and cyanobacterium, or between all three organisms, carries the allies into the most hostile places on land. Beginning as single fungal spores that germinate and recruit their photosynthetic partners as the colony expands, or from larger clumps of fungal filaments enclosing some algal cells, lichens make home on the most unpromising surfaces of rocks, cut stone, and concrete. They thrive on intermittent rations of moisture and inorganic nutrients, and survive exposure to intense sunlight and ultraviolet radiation, and big swings in temperature. By living inside porous rocks, *cryptoendolithic* lichens enjoy more stable environmental conditions than the surface symbioses, and obtain some protection from ultraviolet light. In the Dry Valleys of the Antarctic, sandstones are marked by three or four bands with different colors: the uppermost black speckled zone is colonized by lichenized fungi with heavily melanized cells; gaining UV protection from the melanized cells, fungi in the succeeding white layer lack pigmentation; next deepest is a mixture of the green algae lacking their fungal associates and cyanobacteria, and, finally, 10 millimeters beneath the crust, cyanobacteria are the only residents of the blue-green layer.[51]

The hardiness of lichens has been tested in space. Three common lichen species survived spaceflights of ten days or more on experimental platforms attached to unmanned Russian satellites. Dried by the vacuum of space and exposed to the full glare of electromagnetic radiation from the sun, they hibernated during these missions and resumed photosynthetic activity a couple of days after their return to earth.[52] Intact lichens survived a few months of exposure to space simulations in the lab in which the symbiosis was subjected to vacuum and ultraviolet radiation. Fungus and alga didn't fare as well when they were separated, although a few of the algal cells survived if they were buried inside cell clusters.[53] Spores of the bacterium *Bacillus subtilis* showed greater endurance than the eukaryotes, germinating after six years' exposure to space on a NASA module that completed more than 32,000 orbits of earth.[54]

The existence of cryptoendolithic lichens and the survival of microbes in space encourage champions of Panspermia as a mechanism of biosphere initiation. This hypothesis suggests that extraterrestrial life is common in the universe and is dispersed between planets by asteroids. It doesn't address the most fundamental problem in biology, of course, which is how life originates *anywhere* in the first place.

8

New Jerusalem

The World was all before them, where to choose
Thir place of rest, and Providence thir guide:
They hand in hand with wandring steps and slow,
Through Eden took thir solitarie way.
—*Milton, Paradise Lost, Book XII*

Paradise
Is exactly like
Where you are right now
Only much much
Better.
—*Laurie Anderson*

History, according to the heartfelt judgment of a young scholar in Alan Bennet's marvelous play, *The History Boys*, "[is] just one fuckin' thing after another." Biology is a bit like this, because evolution works with a single set of raw materials within the constraints imposed by the planet's environmental conditions. Every cell is surrounded by a lipid membrane, encodes its information in nucleic acids, and manufactures proteins. Every cell is powered like a battery and uses the electrical current carried by ions to import food, signal to its neighbors, and excrete waste products. At the same time, birds are different from bacteria because biomolecules can be arranged in an immense number of combinations, and sufficient time has elapsed to shuffle the molecules to suit every accommodation on earth. There is a tussle here, between the picture of life drawn from the viewpoint of thermodynamics and the experience of nature afforded anyone walking in the woods or looking at a compost heap with a hand lens. Both views are correct: sameness at the level of life's essence, variety in its manifestation.

Human comprehension of biology has always been distorted by our innate occupation with organisms that are roughly the same size as us, and scientists have believed, until very recently, that organisms of our size are the most important ones for understanding life. Until the seventeenth century, the obvious impediment was our blindness to things smaller than fleas. The slight magnification of nature by Galileo's friends at the Accademia dei Lincei—no more than a well-made hand lens can show us today—was nonetheless revelatory and soon, with the evolution of the microscope, the universe of microorganisms was laid bare. Prospects for intellectual recalibration began with these inventions, but the microscopic didn't bleed into popular consciousness until the link between germs and disease was established in the nineteenth century. During my lifetime we have learned that a far greater repository of biological diversity exists among the unicellular organisms and the viruses than we find throughout the animal and plant kingdoms. Yet, even in the twenty-first century the majority of professional scientists are preoccupied with macrobiology. This is a problem for science and for our species.

Ecologists have exemplified this tension between the macro and the micro of biology. For more than 60 years, ecologists have been interested in understanding how the biodiversity within different ecosystems is determined. Throughout the twentieth century, the number of plant and animal species was viewed as the primary metric of biodiversity. Investigators identified a number of variables that influenced species richness, including climate, the heterogeneity of habitats within the ecosystem, and the abundance of solar radiation. Rainforests support lots of species because their climate is relatively uniform throughout the year, the trees and shrubs create an abundance of distinct habitats, and the sun shines year round. The stability of the ecosystem is another significant consideration. Some tropical forests are so old that evolution has had time to birth many of their younger species.

Contemporary ecologists continue to study these questions, but they recognize now that species richness is only one of multiple measures of biodiversity.[1] Defining the spatial scale of the analysis is important. Some scientists are interested in patterns of diversity within a particular ecosystem, comparing the tree canopy with the shrub layer of a forest, others are concerned with a finer scale, studying differences between neighboring patches in grasslands.

It is evident that ecosystems which support greater numbers of species are also the most productive. The mechanistic link between diversity and productivity is another of the key questions in ecology. The most compelling explanation is that different species make use of different sets of resources so that an assortment of plants has a harmonizing effect, maximizing photosynthesis in a given area of grassland or forest. This is called *niche complementarity*. The concept is complicated by the realization that the range of plant types, rather than the number of species, is a stronger determinant of stability and productivity than species richness alone. Sedges, for example, are better at growing in waterlogged soils than grasses, which means that a mixture of sedges and grasses may be more productive than the richest assortment of grasses.

For a long time, plant ecologists looked at the number of plant species as well as the distribution of different types of plants in particular settings, and developed models of productivity to explain how ecosystems worked. Animal ecologists, on the other hand, pursued similar ideas about animal diversity, and a few of the more interdisciplinary researchers blended these concerns by looking at the effects of herbivores on plant productivity. Microorganisms were included in the standard models of nutrient flow, with fungi, for example, listed as decomposers in models of the carbon cycle. The emphasis, however, was always on plants and animals. Until quite recently, plant and animal ecologists ignored microbes. Microbial ecology was a separate and specialized endeavor.

This is a broad-brushstrokes picture of ecology, but few ecologists of my generation will dispute the contention that the zeitgeist has changed in the last 20 years. It is no longer permissible for a plant ecologist to ignore soil microorganisms: there is little likelihood of receiving funding for experiments on an invasive plant species that makes no effort to examine the diversity of associated mycorrhizal fungi. Ecology cannot be taught any more without considering the importance of microorganisms, and this is a very good thing indeed. By introducing microbes into models of terrestrial ecosystems, investigators have found that fungi and bacteria are actually driving plant productivity.[2] In other words, ecologists had omitted the most important players in their models of ecosystem function. A potent mechanism at work here is the role of fungi in plant disease. As plant diversity decreases, the impact of a single pathogen becomes amplified. This is obvious in the case of monoculture agriculture: if a wheat field is attacked by a rust, crop productivity falls and there are no other plants to take up the slack. The same sort of thing happens in natural ecosystems. The impact of a single pathogenic fungus tends to be muted if plant diversity is high. This is a matter of common sense.

Interactions between plants and microbes go well beyond the effects of pathogens, of course, and studies show that harmless soil microbes impact plant productivity through their influence upon nutrient availability. One approach to studying the details of these symbioses is to control them by growing the plants in sterile greenhouse facilities in containers, or *microcosms*, filled with soil mixtures inoculated with particular fungi and bacteria. Microcosm studies are unambiguous in demonstrating the reliance of the majority of plant species on fungi that establish mycorrhizae. Complementary *macrocosm* experiments, in which field plots are inoculated with fungi, show that the number of plant species tracks the number of fungi.[3] Finally, most plants do not thrive in microbe-free soils, and the usual relationship between plant diversity and productivity collapses without bacteria.[4] The outcomes of these experiments are reminiscent of the observed debilitation of germ-free, or gnotobiotic, mice. Ecosystems, like individual animals, don't work very well without microbes.

Microbes are also imposing themselves upon more specialized fields of ecology. *Restoration ecology* concerns the rehabilitation of damaged ecosystems and contaminated soils. Plants were the sole focus of this inquiry a decade or so ago. Today, investigations on mycorrhizal fungi in these habitats and their interactions with plants are de rigueur. *Conservation biology* has lagged behind restoration ecology in this respect and remains fixated upon macrobiology. When microorganisms are included in conservation studies they appear only as disease agents of larger, *more interesting* species.[5] The fungal pathogen of amphibians has made it to primetime in the major journals concerned with conservation, and microbial pathogens of trees wax and wane according to the current newsworthy epidemic—Dutch elm disease, sudden oak death, ash dieback. The lack of interest in the rest of the news about microorganisms, which is almost the entire story, is almost comedic.

Some of this neglectfulness can be pardoned by the fact that until the advent of metagenomic technologies microbial ecologists had too little to contribute to the topic of conservation. There wasn't enough information on the bacteria in soils to say anything meaningful about their importance in plant diversity. Nevertheless, all biologists have known for decades, if they thought about it, that microbes are more important than frogs in maintaining a biosphere capable of supporting humans. Tom Curtis championed the microbiological view of ecology with the following provocation:

> If the last blue whale choked to death on the last panda, it would be disastrous but not the end of the world. But if we accidentally poisoned the last two species of ammonia-oxidizers, it would be another matter. It could be happening now and we wouldn't even know.[6]

Whales, as I explained earlier, have little effect on nutrient cycling compared with the energy fluxes through marine microbes. This fact has no bearing upon the human relationship with whales, or the importance of whales to themselves and to the organisms that they consume and those that consume them. Watching a breaching humpback whale off Cape Cod and hearing the massive exhalation through its paired blow holes, it seems clear that the mammal is running the show. A microscope and some imagination is required to relegate the whale to the background and absorb the fact of the microbial hegemony in the gray Atlantic water and everywhere else.

By adding microbes to the public discourse we may get closer to comprehending the real workings of the biosphere and the growing threat to their perpetuation. Interest and indifference to conserving different species shows an extraordinary bias in favor of animals with juvenile facial features, "warm" coloration, "endearing" behavior (fur helps too), and other characteristics that appeal to our innate and cultural preferences. The level of discrimination is surprising. Lion cubs have almost universal appeal, and it must take a lifetime of horrors to numb someone to the charms of a baby orangutan. But we make subconscious rankings of animals of every stripe. Among penguins, for example, we prefer species with bright yellow or red feathers.[7] The charismatic megafauna are very distracting, and the popularization of microbial beauty will require a shift in thinking, a subtlety of news coverage, a new genre of wildlife documentary. The ethical responsibility lies with the nations that are engaged in modern biology.

In the first chapter I mentioned the quixotic proposal to catalog every species championed by the Harvard biologist E. O. Wilson. Interest in this futile task continues, with calls from other prominent scientists for naming species "before they go extinct."[8] A cheerful projection suggests that a catalog of five million or so species could be completed within 50 years at a cost of around

US$1 billion per year. This is less than 2 percent of the annual federal investment in scientific research in the United States. The authors of this estimate argue that by naming things we might be in a better position to curb their annihilation. Is this sensible? Name recognition isn't a big problem for tigers and rhinoceros. More logical justifications for this taxonomic marathon speak to the fundamental importance of identifying something as a species to enable the proper exploration of biodiversity, and another stimulus is that the inventory would help determine rates of extinction. There is a strain of desperation here. Today's biologists working on this encyclopedia would become co-authors of a holy book of sorts, a Testament of Ignominy, against which future generations could gauge how much damage we did.

An obvious shortfall of this proposal, as its proponents would agree, is that it wouldn't tell us anything about microorganisms. And that is a tremendous problem and one of the stumbling blocks to accepting that a 50-year taxonomic exercise is worth funding. Biologists, as a community, are still finding it difficult to emerge from the stamp-collecting stage of our science. Whether we are talking about molecular methods or dried sheets in herbaria and drawers filled with disemboweled birds, the importance of the taxonomic exercise deserves some objective analysis. Physics did not stop after Newton; why did so much of biology conclude "Mission Accomplished" after Darwin?

If extinction is the thing we are trying to forestall, we would be better placed in trying to save habitats. The inhabitants of threatened forests would tend to come along automatically, subject to the usual problems with poaching in the remaining wildish places. Because animals and their onboard microbes live in specific habitats, and the habitat is defined, to a large degree, by its plants and the soil microbiome, saving a forest can conserve a lot of things without our ever knowing that they are there.

Greater appreciation of the microbial isn't guaranteed to change the study of biology. The application of metagenomics has already resulted in a groundshift in the science, but there are a lot of uncertainties about the most fruitful ways to proceed. The sheer size of microbial populations suggests that current sampling methods may be inadequate to the task of assessing genetic diversity. A DNA library of 1,000 clones sounds impressive, but if this was amplified from the community of more than one trillion microbes in the nutrient-rich water of my small pond we would have sampled fewer than one in one billion of its cells.[9] Some comfort is found in the annual increase in sequencing speeds and decrease in costs per sequence. The most ambitious experiments on the gut microbiome have analyzed 10,000 or more clones from single samples, and yet the potential shortcomings of the molecular exploration remain. The diversity of a microbial community increases with the detail and depth of the analysis; the more we probe, the more we see. One trillion is an awful lot of individual cells.

Another challenge for biologists trying to understand the activities of the smallest organisms is that most of us are unaccustomed to thinking about the spatial scale of the environment that matters to a single cell.[10] Each of the planktonic bacteria in my pond has a unique life experience shaped by fluctuations in the availability of dissolved ions, changes in temperature and light intensity, contact with other bacteria, and attack by viruses. Gene expression inside the tiny cells is adjusted to maintain energy production and maximize the prospects for cell division. Motile cells with spinning flagella navigate the pond water, responding to gradients in dissolved oxygen and organic nutrients as well as local clouds of metabolites secreted by eukaryotes. Feces puffed from fish add pulses of organic matter to the pond and drops of tree sap plummet through the water column leaving tails of syrup like tiny comets. The pond is a mosaic of microbes and their food. Bacteria lacking flagellar motors are moved by ripples from the pond pump and the flicking of fish tails; convection currents circulate the water too, bringing colder water from the bottom toward the surface warmed by sunbeams; a rain shower cools and mixes the surface water and the belly flop of a frog is the microbial equivalent of an asteroid strike.

Besides the planktonic bacteria, prokaryotes swarm in the silt and over the plastic surface of the pond liner. Bacteria coat algal filaments and the leaves of plants dangling in the pond. Other microbes fill the guts of the fish, frogs, and invertebrates. Every community displays a different version of behavioral complexity. Quorum sensing allows a population of cells to gauge its density and coordinate gene expression. Bacteria secrete signal molecules called *autoinducers* that diffuse throughout the colony, and the concentration of these chemicals serves as an accurate proxy for the number of collaborating cells. When the level of this compound reaches a particular threshold it triggers community-building activities like the dense packing and adhesion of cells to create a protective *biofilm*. In other conditions, quorum sensing activates the invasive growth of pathogens, spore formation, and bioluminescence.

The range of activities of individual microorganisms added to the genetic and physiological diversity of bacteria complicates the goal of developing a holistic description of any ecosystem. Macrobiology is knotty enough on its own, but the introduction of the microscopic can drive models of ecosystem dynamics to spec-tacular levels of incomprehensibility without enhancing their predictive power. The *Amoeba in the Room* began with a quest to comprehend the diversity of organisms in my pond. This relied, for the most part, upon the use of the microscope. The analysis of the sea, soil, air, human gut, and extreme environments in subse-quent chapters rested more upon molecular methods than visual interrogation, yet even the latest technology is lacking. If the life in my pond and my colon are beyond comprehension, what hope do we have of understanding how the open ocean works?

Progress might be accelerated by changing the culture of biology to emphasize the micro over the macro. This could be a game changer for the science, but requires a major shift in the way we teach biology. As a veteran higher educator, I have some cre-dentials for saying that most of us have made a big mess in con-vincing lots of people that biology is a spectacular subject deserving their deep and lifelong engagement. I am not sure how we have failed, or how we could be doing a better job, but here's an example of the problem. The point of the customary micros-copy class at the beginning of a biology lab course is not—and should never have been—to learn to use the microscope. It is an entrée to the student's inquiry into the nature of life. The act of putting a drop of fluid on a microscope slide and viewing it at up to 1,000 times its actual size can be an awe-inspiring experience, no less a thrill than looking at the night sky with a telescope or binoculars.

Microscopes and telescopes make the invisible visible: the night sky dotted with a few weak stars becomes an endless shower of light; a cloudy drop of pond water is filled with spin-ning, whirling, and gliding cells. There is something rotten in the state of Denmark, and everywhere else, when a student yawns when introduced to the microscope. A common exercise in this class is to have the eager scholars swab their mouths and view their globs of mucilage on glass slides. Among the debris they can see buccal cells rubbed from the epidermis on the inside of their cheeks. These are large, flattened cells with prominent nuclei, looking like fried eggs because—which is worth reminding students—chicken eggs are single cells too. The exercise, if approached without cynicism, should counter ennui. Look at that beautiful cell: some of the fine granules are mitochondria that you inherited from your mother; the nucleus houses your 46 chromosomes; you are looking at the stuff of yourself; there is nothing more to you, nor anything less, than what you spy through these eyepieces. The fact that this experience is so often underwhelming is an educational heartbreak, or, at least, an opportunity missed.

The curriculum in the majority of degree programs in the biological sciences emphasizes human biology and the biology of our close vertebrate relatives. The same is true of high-school curricula, and the prominence given to hairy, feathery, and scaly things has always been a fact of science education. In addition to the biomolecules, metabolism and physiology, and genetics and evolution, most introductory biology courses at colleges and universities include a section on biodiversity in which the characteristics of different groups of organisms are described. The time dedicated to each slice of life varies according to the textbook used in the course and the interests of the professor. The fungi may get one lecture if they are lucky. Another class is dedicated to the bacteria, and, often, archaea and viruses are bundled into the same session. Exceptions abound, but this balance of topics parallels the coverage of biology in the best-selling introductory texts on college biology. The organismal part of the course is a gross misrepresentation of the facts of life.

In the Preface I referred to plants as vehicles for their cyanobacterial chloroplasts. Pursuing this metaphorical line, there is some merit in thinking about "The Selfish Bacterium" as an analog to Richard Dawkins' popularization of the gene as the element of continuity throughout the history of life. Humans, for familiar example, can be regarded as temporary conduits for primate genes, as carriers of an immense repository of prokaryote and viral instructions, or as shills for the transportation and replication of bacterial mitochondria. All of these representations have some scientific validity. None of them affect the preoccupation of the individual with everyday concerns—familial, financial, and so on.

For some people the scientific deconstruction of the body has a profound effect upon tolerance for the vagaries of religious doctrine. Deep engagement in the principles of Darwinian evolution has shaken, if not abolished, the faith of many people in supernatural ideas about the special place of *Homo sapiens* in a grand scheme. (I have a colleague whose curiosity about intelligent design survived Darwin, only to be crushed when he learned about the endosymbiotic origin of the eukaryote cell.) Even then, the agnostic biologist spends more of the time worrying about her daughter's dental appointment than she does reveling in the fact that everyone in her family is energized by bacterial proteins in the inner membranes of their mitochondria.

Knowledge of the gut microbiome changes the balance a little. Our highly bacterial nature seems significant to me in an emotional sense. I'm captivated by the revelation that my breakfast feeds the 100 trillion bacteria and archaea in my colon, and that they feed me with short-chain fatty acids. I'm thrilled by the fact that I am farmed by my microbes as much as I cultivate them, that bacteria modulate my physical and mental well-being, and that my microbes are programmed to eat me from the inside out as soon as my heart stops delivering oxygenated blood to my gut. My bacteria will die too, but only following a very fatty last supper. It is tempting to say that the gut microbiome lives and dies with us, but this distinction between organisms is inadequate: our lives are inseparable from the get-go. The more we learn about the theater of our peristaltic cylinder, the more we lose the illusion of control. We carry the microbes around and feed them; they deliver the power that allows us to do so.

Viewed with some philosophical introspection, microbial biology should stimulate a feeling of uneasiness about the meaning of our species and the importance of the individual. But there is boundless opportunity to feel elevated by this science. There are worse fates than to be our kind of farmed animal. In his fascinating book, The Limits of Self, French philosopher Thomas Pradeu examined the ramifications of modern immunological theory on the concept of the individual.[11] Much of his argument hinges on the ways in which our microbiome transforms us into chimeric organisms whose functions are integrated by the immune system.

More than 30 years ago, approaching the British equivalent of high-school graduation, I often escaped the school with a girlfriend and wandered around a community garden. Do not imagine Hogwarts for one second; and, worse, ours was a mostly miserable companionship sustained only by the certainty that this was a teenage misfortune from which the future promised deliverance. We called our refuge what we thought it did not resemble: The Garden of Eden. This triangle of grass was ringed by spindly trees that attracted chattering sparrows, and the birds drew the attentions of grimy cats; old men with gray faces shuffled around the garden too, smoking cigarettes while their dogs exercised; candy wrappers and empty bottles decorated the grass. It was an ugly little place at the bottom of a slope beneath a busy road. Calling this Eden was our satire upon the dearth of beauty in our lives.

In this twenty-first century, I have my Ohio woods in spring, washed with the colors of flowers and animated by the buzz of pollinating insects. Some of the apparent differences between the community garden and the Midwestern woods are an illusion. The beauty of a forest imposes itself on us through the look, smell, sound, and feel of its plants and animals. Its wider significance—the activities that sustain humanity—lies elsewhere, in the functioning of an intact ecosystem and its power to cleanse the air and purify the groundwater. This, like the microorganisms that perform much of the work, is invisible. The evident sensuality of the forest as well as its hidden functions are both important things from our perspective; important because the woods have the power to elevate our feelings, boost our mood, and because without them, our species cannot prosper.

We will have a better feel for the value of the big organisms as well as the ecosystems they inhabit when more of them are gone. But this book is about the little guys. Beyond celebrating the microscopic nature of most of nature, I have suggested that places like my teenage anti-Eden harbor a lot more life than we ever imagined, that there is so much more to life than the portion of it which is immediately apparent. Those teenagers holding hands on the green triangle did not know that they had named their haven so aptly. They did not know that all amid them stood the Tree of Life.

Notes

PREFACE

1. August Johann Rösenhof (1705–1759), a German naturalist and painter of exquisite insect miniatures, described and illustrated *Amoeba* in his *Insecten-Belustigung (Recreation among Insects)* in 1755. Context is provided by J. Leidy, *The American Naturalist* 12, 235–8 (1878).

CHAPTER 1

1. D. L. Meyer and R. A. Davis, *A Sea Without Fish: Life in the Ordovician Sea of the Cincinnati Region* (Bloomington, IN: Indiana University Press, 2009).
2. J. M. Scamardella, *International Microbiology* 2, 207–21 (1999).
3. The names Archaea and Bacteria are capitalized when making an explicit reference to these prokaryotes as taxonomic groups (Domains). Capitalization is dropped in common usage.
4. The formal names of the supergroups are as follows: Amoebozoa, Hacrobia, Stramenopila, Alveolata, Rhizaria, Archaeplastida, Excavata, and Opisthokonta. The stramenopiles, alveolates, and rhizarians are relatively closely related and are treated as a single supergroup baptized SAR in some schemes of classification. The authors of a recent classification of the eukaryotes split the hacrobians into a number of individual groups because evidence for their evolutionary relatedness is slim. The hacrobians don't hang together as tightly as the other supergroups. See S. M. Adl et al., *The Journal of Eukaryotic Microbiology* 59, 429–93 (2012). There is a lot of subjectivity involved in this kind of brute taxonomic carving of the eukaryotes. I'm retaining the name hacrobian for the sake of narrative simplicity and pending additional data from phylogenetic research.
5. L. W. Parfrey, D. J. G. Lahr, and L. A. Katz, *Molecular Biology and Evolution* 25, 787–94 (2008).
6. C. L. McGrath and L. A. Katz, *Trends in Ecology and Evolution* 19, 32–8 (2004).
7. J. Leidy, *U.S. Geological Survey of the Territories Report* 12, 1–324 (1879); J. O. Corliss, *Protist* 152, 69–85 (2001).
8. E. O. Wilson, *Trends in Ecology and Evolution* 18, 77–80 (2003); <http://www.eol.org>

NOTES

9. H. D. Thoreau, *Walden, or Life in the Woods* (Boston: Ticknor and Fields, 1854).
10. <http://www.skepticwonder.fieldofscience.com/>
11. For more on the biology of oomycete water molds see N. P. Money, *Mr. Bloomfield's Orchard: The Mysterious World of Mushroom, Molds, and Mycologists* (New York: Oxford University Press, 2002).
12. G. W. Beakes and S. L. Glockling, *Fungal Genetics and Biology* 24, 45–68 (1998).
13. T. Fenchel, *Protist* 152, 329–38 (2001). For more general information on dinoflagellates see J. D. Hackett et al., *American Journal of Botany* 91, 1523–34 (2004).
14. The intestinal parasite *Giardia* is an example of a non-photosynthetic excavate.

CHAPTER 2

1. D. C. Lindberg, *Isis* 58, 321–41 (1967).
2. John Wedderburn (1583–1651), a disciple of Galileo, said that the Pisan scientist was already using his microscope in 1610. In his *Il Saggiatore* (*The Assayer*), published in 1623, Galileo referred to a "telescope modified to see things very close." This suggests that the early telescope *was* the first microscope. The method of modification isn't clear. One possibility is that he extended one of his smaller telescopes to permit magnification of specimens placed more than three feet from the objective lens. This wouldn't have been very practical and it seems likely that he would have recognized that comparable magnification could be achieved with a single convex lens. This hypothesis is discussed by R. S. Clay and T. H. Court in *The History of the Telescope* (London: Charles Griffin and Company, 1932), and is consistent with a description in a letter written by French astronomer, Nicholas-Claude Fabri de Peiresc, in 1624, referring to Galileo's demonstration of a magnifying instrument that "was of no greater height than a dining-room table." By the 1620s, Galileo had made microscopes with shorter tubes for use on a table top. A superb treatment of the early history of microscopy is given in a classic book by W. B. Carpenter, *The Microscope and its Revelations*, 8th edition revised by W. H. Dallinger (London: J. & A. Churchill, 1901).
3. The quote comes from de Peiresc's 1624 letter (n. 2).
4. There are many competing views on this point of scientific history. One version of the story suggests that Galileo discovered the close-up magnifying effect of the telescope in 1610, yet made no attempts to fabricate a dedicated compound microscope until the 1620s after he had seen a Dutch instrument made by Cornelis Drebbel.

5. The 1625 study of bees was published in three parts: a single engraved sheet of illustrations titled, *Melissographia*; a 90-line poetic examination of bees, *Apis Dianiae*, and a huge broadsheet, *Apiarium*, that described the magnified features of the insect, gathered all manner of scholarship on bees, and lavished praise upon Pope Urban VIII. The Barberini coat of arms appears at the top of *Apiarium*, crowned by papal crown and keys. D. Freedberg, *The Eye of the Lynx: Galileo, His Friends, and the Beginnings of Modern Natural History* (Chicago: The University of Chicago Press, 2002); C. S. Ball, *Proceedings of the Oklahoma Academy of Science, Section D, Social Sciences* 46, 148–51 (1966). Francesco Stelluti published an improved engraving of bee anatomy in a peculiar supplement to a translation of Latin poetry: *Persio Tradotto* (1630).

6. G. Galilei, *Selected Writings*, translated by W. R. Shea and M. Davie (Oxford: Oxford University Press, 2012). The device of the debate between advocates of different viewpoints was also used to brilliant effect by David Hume in his *Dialogues Concerning Natural Religion*, published posthumously and anonymously in 1779. Hume's *Dialogues* was a forceful critique of arguments favoring the existence of God.

7. R. Hooke, *Micrographia or Some Physiological Descriptions of Minute Bodies, Made by Magnifying Glasses; with Observations and Inquiries Thereupon* (London: J. Martyn and J. Allestry, 1665).

8. J. Aubrey, *Brief Lives* (London: Penguin Books, 2000).

9. T. Shadwell, *The Virtuoso* (Lincoln and London: University of Nebraska Press, 1966).

10. Shadwell, n. 9, act I, scene ii, lines 7–10. The next lines in the play belong to Gimcrack's other niece, Miranda, who describes her uncle as, "One who has broken his brains about the nature of maggots, who has studied these twenty years to find out the several sorts of spiders, and never cares for understanding mankind" (I.ii.11–13).

11. *The Diary of Robert Hooke M.A., M.D., F.R.S. (1672–1680)*, edited by H. W. Robinson and W. Adams (London, 1935).

12. C. Dobell, *Antony van Leeuwenhoek and His "Little Animals"* (New York: Russell & Russell, 1958); E. G. Ruestow, *The Microscope in the Dutch Republic: The Shaping of Discovery* (Cambridge: Cambridge University Press, 1996). Nine years after his first description of the bacteria on his teeth, Leeuwenhoek took another look at his teeth scrapings and was surprised to find nothing. Believing his teeth to be "uncommon clean," he probed around the back of his mouth and was rewarded with the sight of, "an inconceivably great number of little animalcules." After much speculation on the subject, he concluded that their absence on his "frontmost grinders" was

due to his practice of drinking morning coffee, "as hot as I can, so hot that it puts me in a sweat...the animalcules that are in the white matter on the front teeth...being unable to bear the hotness of the coffee, are thereby killed."

13. B. J. Ford, *Single Lens: The Story of the Simple Microscope* (New York: Harper & Row, 1985).
14. Ruestow, *The Microscope in the Dutch Republic* (n. 12).
15. G. Adams, *Micrographia Illustrata or, The Knowledge of the Microscope Explain'd* (London: Published by the author, 1746). The terms infusoria and animalcules were used to interchangeably in the early literature on microscopy; in later works, infusoria is limited to bacteria and protists, and animalcules applied exclusively to microscopic animals (e.g., rotifers). Carpenter, *The Microscope and its Revelations* (n. 2), discusses this etymology in some detail.
16. P. Micheli, *Nova Plantarum Genera, Iuxta Tournefortii Methodum Disposita* (Florence: Bernardi Paperinii, 1729).
17. D. L. Hawksworth, introduction to P. Micheli, *Nova Plantarum Genera, Iuxta Tournefortii Methodum Disposita* (Richmond: The Richmond Publishing Co., 1976).
18. J. R. Baker, *Abraham Trembley of Geneva: Scientist and Philosopher 1710–1784* (London: Edward Arnold, 1952); M. J. Ratcliff, *Isis* 95, 555–75 (2004).
19. A. Trembley, *Mémoires, Pour Servier à L'Histoire d'un Genre de Polypes d'eau Douce, à Bras en Forme de Cornes* (Leiden, the Netherlands: Jean and Herman Verbeek, 1744); S. G. Lenhoff and H. M. Lenhoff, *Hydra and the Birth of Experimental Biology—1744. Abraham Trembley's Mémoires Concerning the Polyps* (Pacific Grove, CA: The Boxwood Press, 1986).
20. M. W. Wartofsky, *Diderot Studies* 2, 279–329 (1952).
21. V. P. Dawson, *Nature's Enigma: The Problem of the Polyp in the Letters of Bonnet, Trembley and Réaumur* (Philadelphia: American Philosophical Society, 1987).
22. Anonymous, *Female Inconstancy Display'd in Three Diverting Histories, Describing the Levity of the Fair Sex*, 2nd edition (London: Thomas Johnson, 1732).
23. S. Centlivre, *The Basset Table*, edited by J. Milling (Toronto: Broadview Editions, 2009).
24. T. Chico, *Comparative Drama* 42, 29–49 (2008).
25. H. Baker, *Of Microscopes and the Discoveries Made Thereby*, in two volumes (London: R. Dodsley, 1742). The quotation on *Volvox* comes from Vol. II, *Employment for the Microscope*. Marveling upon the complexity of

NOTES

microscopic organisms, Baker wrote (Vol. II, p. 30), "Perfection appears every where; that Minuteness is no Mark of Meanness...an Atom to Omnipotence is as a World, and a World but as an Atom; in the same Manner as to Eternity one Day is as a thousand Years, and a thousand Years but as one Day." It seems unlikely, to me at least, that William Blake penned the famous lines following "To see a World in a Grain of Sand" in *Auguries of Innocence* (written in 1803, published in 1863), without Baker's work in mind, albeit subconsciously.

26. *Amoeba* was discovered by August Johann Rösel von Rosenhof in the 1750s. Von Rosenhof was a German miniature painter who specialized in studies of insects. J. Leidy, *The American Naturalist* 12, 235–8 (1878).

27. G. C. Ainsworth, *Introduction to the History of Mycology* (Cambridge: Cambridge University Press, 1976).

28. Carpenter, *The Microscope and its Revelations* (n. 2).

29. N. P. Money, *The Triumph of the Fungi: A Rotten History* (New York: Oxford University Press, 2007).

30. A. de Bary, *Morphologie und Physiologie der Pilze, Flechten, und Myxomyceten* (Leipzig: W. Engelmann, 1866), and translation: *Comparative Morphology and Biology of the Fungi Mycetozoa and Bacteria*, trans. H. E. F. Garnsey, revised by I. B. Balfour (Oxford: Clarendon Press, 1887); E. Haeckel, *Generelle Morphologie der Organismen, Allgemeine Grundzuge der Organischen Formen-Wissenschaft, Mechanisch Begründet Durch die von Charles Darwin Reformirte Descendez-Theorie* (Berlin: Georg Reimer, 1866); R. J. Richards, *The Tragic Sense of Life: Ernst Haeckel and the Struggle over Evolutionary Thought* (Chicago: The University of Chicago Press, 2008).

31. J. Sapp, *Microbiology and Molecular Biology Reviews* 69, 292–305 (2005).

32. N. P. Money, *Mr. Bloomfield's Orchard. The Mysterious World of Mushrooms, Molds, and Mycologists* (New York: Oxford University Press, 2002); N. P. Money, *Mushroom* (New York: Oxford University Press, 2011).

33. The first usage of the terms *zoology* and *botany* (botanie) are recorded in 1669 and 1696.

34. H. Copeland, *Quarterly Review of Biology* 13, 383–420 (1938). Copeland replaced his *Monera* with the *Mychota* in *American Naturalist* 81, 340–61 (1947). The *Anucleobionta* was another name proposed for the bacteria in the 1940s. Copeland discussed these nomenclatural issues in *The Classification of Lower Organisms* (Palo Alto, CA: Pacific Books, 1956).

35. R. Whittaker, *Science* 163, 150–60 (1969).

36. C. R. Woese and G. E. Fox, *Proceedings of the National Academy of Sciences* 74, 5088–90 (1977).

37. E. Mayr, *PNAS* 96, 9720–3 (1998); J. Sapp, *The New Foundations of Evolution: On the Tree of Life* (New York: Oxford University Press, 2009).

38. S. B. Dobranski, *English Literary Renaissance* 35, 490–506 (2005).

CHAPTER 3

1. C. Greuet, in *The Biology of the Dinoflagellates*, vol. 21, edited by F. J. R. Taylor (Oxford: Blackwell Scientific Publishers, 1987), 119–42; W. J. Gehring, *Journal of Heredity* 96, 171–84 (2005); M. Hoppenrath et al., *BMC Evolutionary Biology* 9, 116 (2009); F. Gómez, P. López-García, and D. Moreira, *Journal of Eukaryotic Microbiology* 56, 440–5 (2009). The technical name for the organelle is *ocelloid*. The ocelloid has two components: the *hyalosome* and the *melanosome*. Hoppenrath et al. describe the hyalosome as a "layered cornea-like structure and lens-like inclusion...bounded at the base by an iris-like constriction rings," and the melanosome as "a highly-ordered and pigmented retina-like body that is separated from the hyalosome by a seawater chamber."

2. F. J. R. Taylor, *Biosystems* 13, 65–108 (1980). Some warnowiid dinoflagellates are equipped with nematocysts, which are explosive "stingers" that are probably employed in prey capture. *Erythropsidinium* lacks nematocysts and engulfs other cells within an extruded portion of its cytoplasm referred to as a stomopod (mouth-like projection) or stomopharyngian complex (mouth- and throat-like projection).

3. D. Francis, *Journal of Experimental Biology* 47, 495–501 (1967). This is a beautiful paper in which the author described his ophthalmic examination of *Nematodinum* cells netted from the end of a pier at the Scripps Institute of Oceanography in La Jolla, California. Francis measured the refractive index of the lens by placing intact dinoflagellates in seawater on glass slides and illuminating them with a microscope lamp. He did the same thing with a single lens extracted from a cell and performed control measurements using polystyrene spheres that matched the size of the natural lens.

4. The melanosome part of the ocelloid may be a modified chloroplast; Greuet, *The Biology of the Dinoflagellates* (n.1). Adapting a chloroplast to do something that it does very well already does not make much sense, but the addition of a lens may boost photosynthetic efficiency. Richard Dawkins offers a vivid telling of the bumpy character of evolutionary change in *Climbing Mount Improbable* (New York: W. W. Norton, 1996).

5. The colossal squid and the giant squid are different species. The colossal squid, *Mesonychoteuthis hamiltoni*, is estimated to reach up to 14 meters in

length; the giant squid, *Architeuthis* (the number of species in the genus is unknown), is *only* 13 meters long.

6. A touching obituary of Professor Frank E. Round (1927–2010) was published by R. M. Crawford, *Protist* 162, 542–4 (2011).

7. J. D. Hackett et al., *American Journal of Botany* 91, 1523–34 (2004).

8. Besides symbioses with corals, this astonishingly communal microbe lives inside jellyfish, anemones, zoanthids (coral-like animals that live within coral reefs), snails, clams, flatworms, sponges, and foraminiferans and radiolarians (rhizarian protists).

9. M. Stat, E. Morris, and R. D. Gates, *PNAS* 105, 9256–61 (2006).

10. B. Groombridge and M. D. Jenkins, *World Atlas of Biodiversity: Earth's Living Resources in the 21st Century* (Berkeley: University of California Press, 2002).

11. F. Partensky, W. R. Hess, and D. Vaulot, *Microbiology and Molecular Biology Reviews* 63, 106–27 (1999). Blooms of the cyanobacterium *Synechococcus* reach comparable cell densities to *Procholorococcus*. The proteobacterium *Roseobacter* is another highly abundant photosynthetic prokaryote. It is most prevalent in coastal waters.

12. One octillion is 10 to the power 27, which is close to the order-of-magnitude estimate for the number of atoms in the human body of 7.0×10^{27} according to <http://www.wolframalpha.com>

13. J. A. Sohm, E. A. Webb, and D. G. Capone, *Nature Reviews Microbiology* 9, 499–508 (2011); B. Bergman et al., *FEMS Microbiology Reviews* 37, 286–302 (2012). *Trichodesmium* blooms were first named sea sawdust by sailors voyaging with Captain James Cook to the South Pacific (1768–1771). Despite their name, *Trichodesmium* and many other cyanobacteria have a red brown rather than blue-green color due to their synthesis of carotenoid and phycoerythrin pigments.

14. L. M. Proctor, editor, *Special Issue on "A Sea of Microbes," Oceanography* 20 (2007); D. M. Karl, *Nature Reviews Microbiology* 5, 759–69 (2007).

15. D. G. Mann, *Phycologia* 38, 437–95 (1999).

16. J. A. Raven, *Biological Reviews* 58, 179–207 (1983).

17. Apologies to phycologists who call the fertilized egg an auxospore. Excellent introductions to diatom biology are provided by J. E. Graham, L. W. Wilcox, and L. E. Graham, *Algae*, 2nd edition (Upper Saddle River, NJ: Prentice Hall, 2007), and R. E. Lee, *Phycology*, 4th edition (Cambridge: Cambridge University Press, 2008). Alberto Amato provides a nice review of diatom sex in *The International Journal of Plant Reproductive Biology* 2, 1–10 (2010).

18. L. R. Brand et al., *Geology* 32, 165–8 (2004).

19. United States Geological Survey Mineral Resources Program: <http://www.minerals.usgs.gov/>

20. The following is quoted from the website published by Answers in Genesis, the Kentucky-based Christian apologetics ministry that operates the infamous Creation Museum: "the Flood of Noah's day produced the right conditions for a 'blooming' production of microorganisms and the chalk's rapid accumulation." Source: <http://www.answersingenesis.org/articles/wog/white-cliffs-dover>

21. A. R. Taylor et al., *European Journal of Phycology* 42, 125–36 (2007).

22. R. Doerffer and J. Fischer, *Journal of Geophysical Research* 99 C4, 7457–66 (1994); C. W. Brown and J. A. Yoder, *Journal of Geophysical Research* 99 C4, 7467–82 (1994).

23. L. Beaufort et al., *Nature* 476, 80–3 (2011), and illuminating commentary by D. A. Hutchins, *Nature* 476, 41–2 (2011). Citations to other studies, including those that have reached very different conclusions, are given in this pair of papers.

24. L. R. Pomeroy, *BioScience* 24, 499–504 (1974).

25. J. Steinbeck, *The Log from the Sea of Cortez* (New York: Viking Press, 1941).

26. This point was made very eloquently by Forest Rohwer and Merry Youle in their book, *Coral Reefs in the Microbial Seas* (Basalt, CO: Plaid Press 2010).

27. J. C. Venter et al., *Science* 304, 66–74 (2004); D. B. Rusch et al., *PLoS Biology* 5(3), e77 (2007); S. J. Williamson et al., *PLoS ONE* 3(1), e1456.

28. M. A. Moran and E. V. Armbrust, in Proctor, *Special Issue on "A Sea of Microbes,"* (n. 14), 47–55; <http://www.jgi.doe.gov/>

29. R. Massana and C. Pedrós-Alió, *Current Opinion in Microbiology* 11, 213–18 (2008).

30. H. S. Yoon et al., *Science* 332, 714–17 (2011). Single-cell genomics has been used to explore uncharted branches of the tree of life referred to as "microbial dark matter" in a remarkable analysis of the genomes of 200 individual prokaryote cells obtained from varied habitats: C. Rinke et al., *Nature* 499, 431–7 (2013).

31. R. Stocker, *Science* 338, 628–33 (2012).

32. F. E. Round, R. M. Crawford, and D. G. Mann, *The Diatoms: Biology and Morphology of the Genera* (Cambridge: Cambridge University Press, 1990).

33. A. E. Walsby, *Microbiological Reviews* 58, 94–144 (1994).

34. M. C. Benfield et al., in Proctor, *Special Issue on "A Sea of Microbes,"* (n. 14), 172–87.

35. E. C. Roberts et al., *Journal of Plankton Research* 33, 603–14 (2011).

36. J. L. Howland, *The Surprising Archaea* (New York: Oxford University Press, 2000). Given the pace of research on archaea, I hope that Professor Howland is working on a second edition of this volume.

37. Initial evidence for nitrification by pelagic archaea came from the identification of an ammonium monoxygenase gene in the shotgun sequencing study by Venter, *Science* (n. 27).

38. Ø. Bergh et al., *Nature* 340, 467–8 (1989).

39. D. L. Kirchman, *Nature* 494, 320–1 (2013); Y. Zhao et al., *Nature* 494, 357–60 (2013).

40. C. A. Suttle, *Nature Reviews Microbiology* 5, 801–12 (2007).

41. The *Prochlorococcus* cell is a sphere with radius 300 nm; its volume is 10^{-19} m^3; at a concentration in seawater of 10^5 mL^{-1} the total cell volume is 10^{-14} m^3; 1 mL is 10^{-6} m^3, which means that the ratio of this cyanobacterial cell to seawater is $10^{-14}/10^{-6} = 10^8$. Treating the average marine virus as a sphere with 30 nm radius translates to a single-particle volume of 10^{-22} m^3; at a concentration of 10^7 mL^{-1} the total particle volume is 10^{-15} m^3, which means that the ratio of virus to seawater is $10^{-15}/10^{-6} = 10^9$. This guesstimate harmonizes with the conclusion that prokaryotes (*Prochlorococcus* plus other bacteria and archaea) represent more than 90 percent of the biomass of ocean ecosystems, with viruses accounting for an additional 5 percent of the totality of marine biology.

42. M. B. Sullivan et al., *PLoS Biology* 3, e144 (2005); N. H. Mann, *PLoS Biology* 3, e182 (2005).

43. M. Breitbart et al., *PNAS* 99, 14250–5 (2002).

44. W. H. Wilson et al., *Science* 309, 1090–2 (2005).

45. The discovery of large double-stranded DNA viruses called mimiviruses (Mimiviridae) was announced by B. La Scola et al., *Science* 299, 2033 (2003). Catherine Mary provides a taste of the controversies and professional battles in her profile of Didier Raoult, senior author of the 2003 paper, in *Science* 335, 1033–5 (2012).

46. D. Arslan et al., *PNAS* 108, 17486–91 (2011) Pandoraviruses, discovered in 2013, are even bigger: N. Phillipe et al., *Science* 341, 281–6 (2013).

47. D. Raoult, *Nature Reviews Microbiology* 7, 616 (2009).

48. A. Jelmert and D. O. Oppen-Berntsen, *Conservation Biology* 10, 653–4 (1996); A. J. Pershing et al., *PLoS ONE* 5, e12444 (2010).

49. M. L. Walser, S. C. Nodvin, and S. Draggan, in *Encyclopedia of Earth*, edited by C. J. Cleveland (Washington, DC: Environmental Information Coalition, National Council for Science and the Environment, 2011), <http://www.eoearth.org/article/Carbon_footprint>

Example of calculation: The average human is responsible for the emission of 4 tons of carbon dioxide (CO_2 equivalents) per year. Dead whales carrying 160,000 tons of carbon to the abyss remove the equivalent of 587,200 tons of carbon dioxide (the ratio of molecular weights is 44/12 = 3.67). This balances the emissions of 146,800 average humans, or 0.002 percent of the global population.

50. K. O. Buesseler et al., *Science* 316, 567–70 (2007).

CHAPTER 4

1. C. Darwin, *The Formation of Vegetable Mould through the Action of Earthworms* (London: John Murray, 1881). Darwin's fascination with worms lasted more than 40 years. The earliest evidence of this occupation derives from his paper on earthworms read at a meeting of the Geological Society of London in 1837: C. Darwin, *Proceedings of the Geological Society of London* 2, 574–6 (1838).

2. U. Kutschera and J. M. Elliott, *Applied and Environmental Soil Science* 2010, (2010).

3. C. R. Darwin, *The Structure and Distribution of Coral Reefs. Being the First Part of the Geology of the Voyage of the Beagle, Under the Command of Capt. Fitzroy, R. N. During the Years 1832 to 1836.* (London: Smith, Elder and Co., 1842).

4. Other biologists have contemplated the limitations of the nineteenth-century view of the tangled bank, including C. Currie, *Microbe Magazine* 6, 440–5 (2011).

5. D. C. Price et al., *Science* 335, 843–7 (2012).

6. J. A. Raven and J. F. Allen, *Genome Biology* 4, 209.1–209.5 (2003). Endosymbiosis was proposed by Russian botanist Konstantin Mereschkowski in 1905. Long before its validation by molecular genetics, striking similarities between bacteria, mitochondria, and chloroplasts had been recognized in electron micrographs. Modern research on the theory originated with Lynn Margulis: L. Sagan, *Journal of Theoretical Biology* 14: 255–74 (1967).

7. W. A. Shear, *Nature* 351, 283–9 (1991); C. K. Keller and B. D. Wood, *Nature* 364, 223–5 (1993); R. J. Horodyski and L. P. Knauth, *Science* 263, 494–8 (1994); A. R. Prave, *Geology* 30, 811–14 (2002).

8. P. Kendrick and P. R. Crane, *Nature* 389, 33–8 (1997); L. A. Lewis and R. M. McCourt, *American Journal of Botany* 91, 1535–56 (2004); S. Wodniok et al., *BMC Evolutionary Biology* 11, (2011).

9. D. L. Royer et al., *American Journal of Botany* 97, 438–45 (2010).

10. Plant collectors and taxonomists have become an endangered species in the twenty-first century: J. Whitfield, *Nature* 484, 436–8 (2012). H. C. J.

Godfray discusses broader challenges for plant and animal taxonomy in *Nature* 417, 17–19 (2002).

11. The Sun is classified as G-type main-sequence star, also known as a yellow dwarf. Energy in its dense core derives from the fusion of hydrogen atoms to produce helium.

12. D. D. Richter and D. Markewitz, *Bioscience* 45, 600–9 (1995).

13. R. Daniel, *Nature Reviews Microbiology* 3, 470–8 (2005).

14. V. Torsvik, J. Goksøyr, and F. L. Daae, *Applied and Environmental Microbiology* 56, 782–7 (1990); J. Gans, M. Wolinsky, and J. Dunbar, *Science* 309, 1387–90 (2005); T. P. Curtis and W. T. Sloan, *Science* 309, 1331–3 (2005). The estimated one million taxa per gram works for a hypothetical soil sample that supports 100,000 individual cells of each of ten dominant species of bacteria plus an average of 99 individuals each of the remaining 999,990 taxa.

15. T. M. Vogel et al., *Nature Reviews Microbiology* 7, 252 (2009).

16. P. C. Baveye, *Nature Reviews Microbiology* 7, 756 (2009).

17. M. T. Madigan et al., *Brock Biology of Microorganisms*, 13th edition (San Francisco, CA: Benjamin Cummings, 2010). One of the stranger metabolic strategies is found in a fast-growing extremophile, *Carboxydothermus hydrogenoformans*, which lives in hot springs on a volcanic island in Russia. This is an example of a bacterium that consumes carbon monoxide and releases hydrogen gas: M. Wu et al., *PLoS Genetics* 1(5), e65 (2005).

18. K. J. van Groenigen, C. Rosenberg, and B. A. Hungate, *Nature* 475, 214–16 (2011).

19. I. M. Brodo, S. D. Sharnoff, and S. Sharnoff, *Lichens of North America* (New Haven, CT: Yale University Press, 2001).

20. N. P. Money, *The Triumph of the Fungi: A Rotten History* (New York: Oxford University Press, 2007).

21. J. Sapp, *Evolution by Association: A History of Symbiosis* (New York: Oxford University Press, 1994).

22. N. P. Money, *Mushroom* (New York: Oxford University Press, 2011).

23. D. Redecker, R. Kodner, and L. E. Graham, *Science* 289, 1920–1 (2000). Liverworts have rhizoids rather than root systems, which disqualifies them from possessing true mycorrhizae (fungus roots). Rhizoids are single-celled threads arranged in a fringe that anchors the bottom of the liverwort to wet soil. The entire surface of the liverwort absorbs water and dissolved nutrients; a root system would be an encumbrance for such a simple terrestrial display of chloroplasts.

24. J. Russell and S. Bulman, *New Phytologist* 165, 567–9 (2005); C. Humphreys et al., *Nature Communications* 1, 103 (2010); A. Jermy, *Nature Reviews Microbiology* 9, 6 (2011). The fungi engaged in these symbioses are members of the

Phylum Glomeromycota. They cannot be cultured separately from their plant hosts. A second group of fungi called the Endogonales may have formed even earlier associations with land plants: M. I. Bidartondo et al., *Biology Letters* 7, 574–7 (2011). These organisms will grow in pure culture.

25. F. K. Sparrow, *Aquatic Phycomycetes (Exclusive of the Saprolegniaceae and Pythium)* (Ann Arbor, MI: University of Michigan Press, 1943). The second edition, published in 1960, is a heftier volume with more than 300 additional pages on the formerly rusticated Saprolegniaceae and *Pythium*. The term phycomycetes, meaning "algal fungi," is obsolete and referred very loosely to fungi and fungus-like microorganisms whose preferred habitats are aquatic.

26. The role of the chytrid in amphibian decline remains controversial and some investigators think that infection by this fungus reflects damage to the animals from other sources. One study concluded that only 14 percent of cases of amphibian decline could be linked to chytrid infection: M. Heard, K. F. Smith, and K. Ripp, *PLoS ONE* 6(8), e23150 (2011).

27. M. J. Powell, *Mycologia* 76, 1039–48 (1984).

28. T. Y. James et al., *Mycologia* 98, 860–71 (2006).

29. E. Lara, D. Moreira, and P. López-Garcia, *Protist* 161, 116–21 (2010).

30. M. D. M. Jones, *Nature* 474, 200–3 (2011).

31. R. W. G. Dennis, *British Cup Fungi and Their Allies: An Introduction to the Ascomycetes* (London: The Ray Society, 1960).

32. K. E. Ashelford, M. J. Day, and J. C. Fry, *Applied and Environmental Microbiology* 69, 285–9 (2003).

33. The website is <http://www.phagesdb.org> and genome sequences are deposited at GenBank (<http://www.ncbi.nlm.nih.gov/genbank/>) which is the annotated collection of all publicly available DNA sequences administered by the National Institutes of Health.

34. F. Rohwer, *Cell* 113, 141 (2003).

35. Prokaryote numbers in unpolluted lakes and rivers average (very roughly) one million cells per milliliter, which is ten times greater than the density of microorganisms in seawater; W. B. Whitman, D. C. Coleman, and W. J. Wiebe, *PNAS* 95, 6578–83 (1998).

36. J. L. Frank, R. A. Coffan, and D. Southworth, *Mycologia* 102, 93–107 (2010).

37. D. C. Sigee, *Freshwater Microbiology* (Chichester: John Wiley & Sons, 2005).

38. R. Logares et al., *Trends in Microbiology* 17, 414–22 (2009).

39. P. A. Sims, D. G. Mann, and L. K. Medlin, *Phycologia* 45, 361–402 (2006).

40. A. J. Alverson, R. K. Jansen, and E. C. Theriot, *Molecular Phylogenetics and Evolution* 45, 193–210 (2007).

CHAPTER 5

1. N. P. Money, Mr. Bloomfield's Orchard: The Mysterious World of Mushrooms, Molds, and Mycologists (New York: Oxford University Press, 2002).

2. The smallest insects are males of the fairyfly species, Dicomorpha echmept-erygis (family Mymaridae): with a length of only 0.14 mm, they are considerably smaller than the single cells of Amoeba proteus. This Costa Rican species parasitizes the eggs of other insects. The fascinating anatomical adaptations that pack the insect nervous system into the tiny bodies of fairyflies are described by A. A. Polilov, Arthropod Structure and Development 41, 29–34 (2012).

3. Lucretius, The Nature of Things, translated by A. E. Stallings (London: Penguin Books, 2007), Book II, quoted phrases from lines 140 and 136. In Book VI the Roman poet speculated that air could be a source of contagious diseases: "when a sky bestirs itself" (1119), "Nature who imports a sky contaminated, a climate that is new to us who, unaccustomed, are more vulnerable to attack" (1135–7).

4. F. C. Meier, The Scientific Monthly 40, 5–20 (1935).

5. R. J. Haskell and H. P. Barss, Phytopathology 29, 293–302 (1939).

6. Carpet Monsters and Killer Spores: The Natural History of Toxic Mold (New York: Oxford University Press, 2004), 148, n. 16.

7. H. E. Schlichting, Air Pollution Control Association Journal 19, 946–51 (1969).

8. D. W. Griffin et al., American Scientist 90, 228–35 (2002), D. W. Griffin, Clinical Microbiology Reviews 20, 459–77 (2007).

9. J. Giles, Nature 434, 816–19 (2005).

10. C. S. Bristow, K. A. Hudson-Edwards, and A. Chappell, Geophysical Research Letters 37, L14807 (2010).

11. C. Darwin, Journal of Researches into the Geology and Natural History of the Various Countries Visited by H.M.S. Beagle, Under the Command of Captain Fitzroy, R. N. from 1832 to 1836. (London: Henry Colburn, 1839); R. S. Cerveny, Bulletin of the American Meteorological Society 86, 1295–301 (2005).

12. The exception is a green alga, Prototheca, which lacks chlorophyll and causes protothecosis. Human cases are very rare and may be associated with impaired immune defenses. Most infections begin with contact through a skin abrasion. B. Leimann et al., Medical Mycology 42, 95–106 (2004).

13. James Salisbury (1823–1905) is best known for his belief in the poisonous nature of vegetables and invention of the Salisbury steak—ground beef flavored with onion, then deep fried or boiled—as the perfect food that should be eaten three times a day. His enthusiasm for meat extended to

his recommended therapy of chopped steak and coffee to control diar-
rhea among his soldiers: J. H. Salisbury, *The Relation of Alimentation and Disease* (New York: J. H. Vail and Company, 1888).

14. J. H. Salisbury, *The American Journal of the Medical Sciences* 51, 51–75 (1866).

15. Malaria is caused by species of *Plasmodium. Plasmodium falciparum* is responsible for 75 percent of all human cases. *Plasmodium* is a genus of apicomplexan protist; the apicomplexans are part of the alveolate super-group that includes the dinoflagellates and ciliates.

16. Lancaster's Starch Factory became the Camp Medill barracks during the Civil War.

17. J. H. Salisbury, *The American Journal of the Medical Sciences* 44, 17–28 (1862).

18. S. Genitsaris, C. A. Kormas, and M. Moustaka-Gouini, *Frontiers in Bioscience* E3, 772–87 (2011).

19. G. Miller, *Science* 313, 428–31 (2006).

20. S. Genitsaris et al., *Frontiers in Bioscience* (n.18).

21. The possible role of toxic cyanobacterial blooms in a cluster of ALS cases in New Hampshire is discussed by T. A. Caller et al., *Amyotrophic Lateral Sclerosis* S.2, 101–8 (2009). More important from the viewpoint of global health issues is the possibility that cyanobacterial toxins may become biomagnified in marine food chains. The widespread distribution of neurotoxic amino acid synthesis among cyanobacteria is reported by P. A. Cox et al., *PNAS* 102, 5074–8 (2010).

22. R. E. Lee, G. J. Warren, and L. V. Gusta, *Biological Ice Nucleation and its Applications* (St. Paul, MN: American Phytopathological Society Press, 2005).

23. B. C. Christner et al., *Science* 319, 1214 (2008).

24. M. O. Andreae and P. J. Crutzen, *Science* 276, 1052–8 (1997).

25. W. D. Hamilton and T. M. Lenton, *Ethology Ecology and Evolution* 10, 1–16 (1998).

26. The authors do not detail the mechanism that would cause changes in wind speed at the surface of the ocean, but suggest that DMS emission would induce convection currents through the release of the latent heat of condensation and that these currents would cause wind gusts.

27. The evolution of wind- and cloud-generation among a population of microbes that facilitated group dispersal would be violated by the development of cheaters that would make use of wind and clouds without investing energy in their production.

28. J. K. M. Brown and M. S. Hovmøller, *Science* 297, 537–41 (2002); D. E. Aylor, *Ecology* 84, 1989–97 (2003).

29. N. P. Money, *The Triumph of the Fungi: A Rotten History* (New York: Oxford University Press, 2007).
30. W. Elbert et al., *Atmospheric Chemistry and Physics* 7, 4569–88 (2007); A. Sesartic and T. N. Dallafior, *Biogeosciences* 8, 1181–92 (2011).
31. For the purpose of mathematical amusement, let's assume that the average-sized fungal spore has a diameter of 10 millionths of one meter (10 micrometers). A microsphere of this size with the same density as water has a mass of 5×10^{-16} tons. Dividing the total mass of airborne spores of 50,000,000 tons by the mass of the single spore, we arrive at a count of 10^{23} spores. Avogadro's number of atoms in a mole is 6×10^{23}, which is close enough to invoke comparison with the global spore cloud.
32. The International Nucelotide Sequence Database (INSD; <http://www.insdc.org>) synchronizes sequence data uploaded to three databases: Genbank, administered by the National Center for Biotechnology Information (NCBI is a branch of the National Institutes of Health), the European Nucleotide Archive, and the DNA Data Bank of Japan.
33. J. Fröhlich-Nowoisky et al., *Biogeosciences* 9, 1125–36 (2011). Molecular analysis of bacteria in the microbiome of the upper troposphere is described by N. DeLeon-Rodriguez et al., *PNAS* 110, 2575–80 (2013).
34. J. Fröhlich-Nowoisky et al., *PNAS* 106, 12814–19 (2009).
35. U. Pöschl et al., *Science* 329, 1513–16 (2010).
36. C. T. Ingold and S. A. Hadland, *New Phytologist* 58, 46–57 (1959). A comparably elegant experiment measured the launch speed of the dung fungus *Pilobolus* using a pair of spinning discs placed in front of its squirt guns. The discs were attached to the same spindle and rotated by a motor. To hit the second disc, the spores had to pass through a hole in the first disc. Most spores were splattered on the first disc, but a few passed through the hole uninterrupted and landed on the second disc. The displacement of the spores on the circumference of the second disc, relative to the aperture in the first, was proportional to their velocity. The launch speed estimated from these experiments was 14 meters per second, or 50 kilometers per hour. E. G. Pringsheim and V. Czurda, *Jahrbücher für Wissenschaftliche Botanik* 66, 683–901 (1927); L. Yafetto et al., *PLoS ONE* 3(9), e3237 (2008). Rather than discharging individual spores, *Pilobolus* launches a sporangium filled with 10,000 spores over a distance of 2.5 meters.
37. The diversity of fungi on herbivore dung is celebrated in beautiful studies by New Zealand investigator Ann Bell. Her books are a celebration of the microscopic wonders that develop on the feces of the marsupial and eutherian mammals and include magnificent water-colored illustrations of

the tiny fruit bodies and the extraordinary spores that they blast into the antipodean air. A. Bell, *Dung Fungi: An Illustrated Guide to Coprophilous Fungi in NewZealand* (Wellington, New Zealand: Victoria University Press, 1983); A. Bell, *An Illustrated Guide to the Coprophilous Ascomycetes of Australia* (Utrecht, The Netherlands: Centraalbureau voor Schimmelcultures, 2005). In the preface to the first book, Ann wrote, "I am aware of the dangers of aiming a book at too wide an audience." This might strike anyone other than experts on dung fungi as an unnecessary concern.

38. O. K. Davis and D. S. Shafer, *Palaeogeography, Palaeoclimatology, Palaeoecology* 237, 40–50 (2006); J. L. Gill et al., *Science* 326, 1100–3 (2009); J. R. Wood et al., *Quarternary Science Reviews* 30, 915–20 (2011).

39. A positive reaction in a skin test is associated with airway constriction when spores bearing the same allergens reach the lungs. Clinical studies have shown allergic sensitivity to *Alternaria* spores in up to 46 percent of asthmatic children. A variety of claims have been made about geographic patterns of *Alternaria* allergy, some suggesting highest sensitivity in desert climates, others highlighting greater prevalence in wetter regions. Among children and adults with severe persistent asthma symptoms, including those admitted to intensive care units for treatment, sensitivity to *Alternaria* is very common. T. W. Lyons, D. B. Wakefield, and M. M. Cloutier, *Annals of Allergy, Asthma and Immunology* 106, 301–7 (2011).

40. C. T. Ingold, *Fungal Spores: Their Liberation and Dispersal* (Oxford: Oxford University Press, 1971).

41. S. Braman, *Chest* 130 suppl., 4S–12S (2006); updated data available on the website for The Global Initiative for Asthma, <http://www.ginasthma.org>

42. J. N. Klironomos et al., *Canadian Journal of Botany* 75, 1670–3 (1997); J. Wolf et al., *Environmental Health Perspectives* 118, 1223–8 (2010).

43. A. S. Amend et al., *PNAS* 107, 13748–53 (2010).

44. G. Krstić, *PLoS ONE* 6(4), e18492 (2011).

45. W. F. Wells, *Airborne Contagion and Air Hygiene* (Cambridge: Harvard University Press, 1955).

46. D. W. Griffin, *Clinical Microbiology Reviews* (n.8).

47. A. A. Imshenetsky, S. V. Lysenko, and G. A. Kazakov, *Applied and Environmental Microbiology* 35, 1–5 (1978). The Soviet researchers reported finding heavily melanized fungal spores in their samples. The presence of UV-protected cells encourages some optimism that living things can survive at the top of the stratosphere, but it would be nice to see some confirmatory research after 30 years. Laboratory simulations show that even the most resistant bacterial spores are destroyed by a few hours of

exposure to the levels of UV radiation encountered in the stratosphere: D. J. Smith et al., *Aerobiologia* 27, 319–32 (2011).

48. D. J. Smith, D. W. Griffin, and A. C. Schuerger, *Aerobiologia* 26, 35–46 (2010). Live fungi and bacteria were isolated from an altitude of 20 kilometers in this study.

49. A. M. Womack, B. J. M. Bohannan, and J. L. Green, *Philosophical Transactions of the Royal Society B* 365, 3645–53 (2010); DeLeon-Rodriguez et al., *PNAS* (n.34).

50. J. Taylor et al., *Philosophical Transactions of the Royal Society B* 361, 1947–63 (2006).

51. B. E. Wolfe and A. Pringle, *ISME Journal* 6, 745–55 (2012).

52. J. Green and J. M. Bohannan, *Trends in Ecology and Evolution* 21, 501–7 (2006); D. Fontaneto, editor, *Biogeography of Microscopic Organisms: Is Everything Small Everywhere?* (Cambridge: Cambridge University Press, 2011).

CHAPTER 6

1. Retroviral genes constitute approximately 8 percent of the human genome; E. S. Lander et al., *Nature* 409, 860–921 (2001). Genes from non-retroviral viruses have also been identified; M. Horie et al., *Nature* 463, 84–7 (2010).

2. A simple fetal microbiome may be started by a trickle of bacteria from the maternal gut into the amniotic fluid; L. J. Funkhouser and S. R. Bordenstein, *PLoS Biology* 11, e1001631 (2013). J. E. Koenig et al., *PNAS* 108, 4578–85 (2011) offer detailed analysis of the development of the infant microbiome.

3. C. A. Lozupone et al., *Nature* 489, 220–30 (2012).

4. G. D. Wu et al., *Science* 334, 105–8 (2011).

5. P. J. Turnbaugh et al., *Cell Host Microbe* 3, 213–23 (2008).

6. R. Ley et al., *Nature* 444, 1022–3 (2006).

7. C. A. Lozupone et al., *Nature* (n. 3).

8. T. C. Hazen et al., *Science* 330, 204–8 (2010).

9. B. L. Cantarel et al., *PLoS ONE* 7, e28742 (2012).

10. There are a variety of approaches to next-generation sequencing. Pyrosequencing detects flashes of light that accompany the addition of complementary nucleotide bases—As, Ts, Gs, and Cs—to single strands of DNA attached to beads bathed in pools of reagents in multiwell plates. This technique mimics what happens when DNA is replicated but renders the reactions visible to the sequencing platform by coupling the addition of each base to the light emission resulting from the oxidation of firefly luciferin. Illumina sequencing is less expensive and reads DNA strands attached to glass slides. Assembly of the complementary strands is tracked from light pulses whose wavelengths are specific to fluorescent

labels attached to each of the four nucleotides. SOLiD is a third technology that utilizes ligation reactions that attach short lengths of DNA (oligonucleotides) rather than single nucleotides to the sequenced strand. The low cost of the latest methods comes at the expense of a limited read length: Illumina provides reads of 100–150 bases and SOLiD is limited to 35 bases, whereas pyrosequencing deals with 400 bases in a single sweep. Many other methods are being developed, including nanopore sequencing in which the sequence of nucleotides is read as a series of electrical pulses as the DNA strand is passed through a channel. T. C. Glenn, *Molecular Ecology Resources* 11, 759–69 (2011); G. M. Weinstock, *Nature* 489, 250–6 (2012).

11. <http://www.genome.gov/sequencingcosts/>
12. W. R. Wikoff et al., *PNAS* 106, 3698–703 (2009); J. K. Nicholson et al., *Science* 336, 1262–7 (2012).
13. M. Balter, *Science* 336, 1246–7 (2012) and websites: <https://www.commonfund.nih.gov/hmp/>; <http://www.metahit.eu/; http://www.genomics.cn/en/index>
14. C. Jenberg et al., *The ISME Journal* 1, 56–66 (2007).
15. L. C. Antunes et al., *Antimicrobial Agents and Chemotherapy* 55, 1494–503 (2011).
16. S. Suerbaum and P. Michetti, *New England Journal of Medicine* 347, 1175–86 (2002).
17. I. C. Arnold et al., *Journal of Clinical Investigation* 121, 3088–93 (2011).
18. B. Linz et al., *Nature* 445, 915–18 (2007).
19. S. Thavagnanam et al., *Clinical and Experimental Allergy* 38, 629–33 (2008).
20. M. G. Dominguez-Bello et al., *PNAS* 107, 11971–5 (2010).
21. M. Kuitunen et al., *Journal of Allergy and Clinical Immunology* 123, 335–41 (2009).
22. N. Cerf-Bensussan and V. Gaboriau-Routhiau, *Nature Reviews Immunology* 10, 735–44 (2010).
23. K. Berer et al., *Nature* 479, 538–42 (2011).
24. I. Martinez et al., *The ISME Journal* 7, 269–80 (2013).
25. E. Van Nood et al., *The New England Journal of Medicine* 368, 407–15 (2013).
26. C. Huttenhower et al., *Nature* 486, 207–14 (2012). I rounded the numbers of samples, sequences, and DNA bases.
27. F. Armougom et al., *PLoS ONE* 4, e7125 (2009). I rounded numbers from this study to single significant figures to arrive at the 40 billion bacteria per gram. For a total fecal mass of 2 kilograms, the gut contains 80 trillion bacteria, which is close to the often-quoted gut microbiome content of 100 trillion cells. This is the highest density of cells recorded for any ecosystem: W. B. Whitman, D. C. Coleman, and W. J. Wiebe, *PNAS* 95, 6578–83 (1998); F. Bäckhead et al., *Science* 307, 1915–20 (2005).

28. H. Zhang et al., *PNAS* 106, 2365–70 (2009); B. Dridi, D. Raoult, and M. Drancourt, *Anaerobe* 17, 56–63 (2011). Some studies are contradictory, suggesting that obesity is associated with lower levels of methanogenic archaea: M. Million et al., *International Journal of Obesity* 36, 817–25 (2012).

29. F. Armougom et al., *PLoS ONE* (n. 27).

30. A. P. A. Oxley et al., *Environmental Microbiology* 12, 2398–410 (2010).

31. Haptophytes and the majority of the cryptomonads are photosynthetic hacrobians; ketablepharids, centrohelid heliozoans, and telonemids are heterotrophic hacrobians (<http://www.tolweb.org/Hacrobia/>). It is possible that species from these groups will be identified in future studies of the gut microbiome.

32. A. Stechmann et al., *Current Biology* 18, 580–5 (2008). Most eukaryotes that occupy low-oxygen or anaerobic habitats possess organelles that are modified mitochondria called hydrogenosomes and mitosomes. The organelle in *Blastocystis* has characteristics of mitochondria and hydrogenosomes.

33. I. Hamad et al., *PLoS ONE* 7, e40888 (2012).

34. I. D. Iliev et al., *Science* 336, 1314–17 (2012).

35. S. Minot et al., *Genome Research* 21, 1616–25 (2012). Phages may complement the immune defenses by controlling bacterial populations in mucus: J. J. Barr et al., *PNAS* 110, 10771–6 (2013).

36. Human genome, initial sequencing by E. S. Lander et al., *Nature* (n. 1); chimp, T. Mikkelsen et al., *Nature* 437, 69–87 (2005); bonobo, K. Prüfer et al., *Nature* 486, 527–31 (2012).

37. R. E. Ley et al., *Science* 320, 1647–51 (2008); R. E. Ley et al., *Nature Reviews Microbiology* 6, 776–88 (2008).

38. Termites and beetle larvae are a notable exception, possessing very complex microbial communities engaged in the anaerobic fermentation of cellulose. Cellulose fermentation is accomplished by two protists in the termite gut, *Trichonympha* and *Mixotrichia*, which are members of the excavate supergroup. The cellulose-degrading enzymes are secreted by endosymbiotic bacteria and the cell surface of the protists is covered with spirochete bacteria whose undulations propel their partners through their fluid surroundings. The surface spirochetes are examples of *ectosymbionts*.

39. C. Huttenhower et al., *Nature* (n. 26).

40. P. W. Lepp et al., *PNAS* 101, 6176–81 (2004).

41. M. A. Ghannoum et al., *PLoS Pathogens* 6, e1000713 (2010).

42. K. Findley et al., *Nature* 498, 367–70 (2013).
43. R. D. Heijtz et al., *PNAS* 108, 3047–52 (2011).
44. As a neurotransmitter, serotonin affects mood, appetite, and sleep, and is a major antidepressant; serotonin in the gut controls peristaltic movement of the intestine. This explains why the release of serotonin by the pathogenic protist *Entamoeba histolytica* causes diarrhea.
45. I shit, therefore I am.

CHAPTER 7

1. T. S. Suryanarayanan et al., *Fungal Biology* 115, 833–8 (2011). Experiments on fungi isolated from the Western Ghats found some species capable of surviving a 2-hour bake at 115 °C, or 239 °F.
2. W. L. Kenney, D. W. DeGroot, and L. A. Holowatz, *Journal of Thermal Biology* 29, 479–85 (2004).
3. E. Blochl et al., *Extremophiles* 1, 14–21 (1997).
4. F. T. Robb and D. S. Clark, *Journal of Molecular Microbiology and Biotechnology* 1, 101–5 (1999); G. N. Somero, *Annual Review of Physiology* 57, 43–68 (1995).
5. K. Kashefi and D. R. Lovley, *Science* 301, 934 (2003).
6. J. A. Mikucki et al., *Science* 324, 397–400 (2009).
7. P. D. Franzmann et al., *International Journal of Systematic Bacteriology* 47, 1068–72 (1997); N. F. W. Saunders et al., *Genome Research* 13, 1580–8 (2003).
8. C. Gerday and N. Glansdorff, editors, *Physiology and Biochemistry of Extremophiles* (Washington, DC: ASM Press, 2007).
9. The oxygen concentration in Lake Vostok is 50 times higher than the surface of a lake at atmospheric pressure. The abundance of damaging oxygen radicals under these conditions is a big problem for cell physiology.
10. <http://blogs.nature.com/news/2012/02/lake-vostok-drilling-success-confirmed.html>. Sequences from more than 1,000 bacteria, 200 or so eukaryotes, and 2 archaea have been amplified from subglacial accretion ice at Lake Vostok: Y. M. Shtarkman et al., *PLoS ONE* 8(7), e67221 (2013). A layer of accretion ice is formed from the uppermost region of lake water beneath the 3.5-kilometer-thick glacier. The presence of microbes in this location bodes well for an active microbiome in the lake below.
11. A. E. Murray et al., *PNAS* 109, 20626–31 (2012).
12. W. B. Whitman, D. C. Coleman, and W. J. Wiebe, *PNAS* 95, 6578–83 (1998).
13. <http://www.iodp.org/Mission/>
14. B. B. Jørgensen and A. Boetius, *Nature Reviews Microbiology* 5, 770–81

(2007); M. A. Lever et al., *Science* 339, 1305–8 (2013). Filamentous bacteria act as electrical cables connecting biochemical reactions at the oxygenated surface with the underlying anoxic sediment via the flow of electrons: C. Pfeffer et al., *Nature* 491, 218–21 (2012).

15. R. Monastersky, *Nature* 492, 163 (2012). A variety of yeasts and filamentous fungi have been identified from their rRNA sequences retrieved from a sediment depth of 50 meters below the Pacific seafloor: W. Orsi, J. F. Biddle, and V. Edgcomd, *PLoS ONE* 8(2), e56335 (2013).

16. S. D'Hondt et al., *Science* 306, 2216–21 (2004).

17. L. Phillips, *Nature News* (May 17, 2012) <http://www.nature.com/news/slo-mo-microbes-extend-the-frontiers-of-life-1.10669>; H. Røy et al., *Science* 336, 922–5 (2012).

18. E. G. Roussel et al., *Science* 320, 1046 (2008).

19. T. Gold, *PNAS* 89, 6045–9 (1992); B. B. Jørgensen, *PNAS* 109, 15976–7 (2012); J. Kallmeyer et al., *PNAS* 109, 16213–16 (2012).

20. A proposed exception to the chemotrophic way of life is a green sulfur bacterium that accomplishes the astonishing feat of conducting photosynthesis in a black smoker vent at a depth of more than 2 kilometers. It is thought to do so by absorbing photons from the flashes of geothermal light: J. T. Beatty et al., *PNAS* 102, 9306–10 (2005).

21. N. Dubilier, C. Bergin, and C. Lott, *Nature Reviews Microbiology* 6, 725–40 (2008).

22. W. Martin and M. J. Russell, *Philosophical Transactions of the Royal Society B* 362, 1887–926 (2007); N. H. Sleep, D. K. Bird, and E. C. Pope, *Philosophical Transactions of the Royal Society B* 366, 2857–69 (2011). Alkaline vent fluid with pH 9–11, would have met an ancient ocean with pH 6 compared with today's pH 8.

23. D. Schulze-Makuch et al., *Astrobiology* 11, 241–58 (2011).

24. J. D. Van Hamme, A. Singh, and O. P. Ward, *Microbiology and Molecular Biology Reviews* 67, 503–49 (2003).

25. C. Schleper et al., *Nature* 375, 741–2 (1995).

26. C. Gerday and N. Glansdorff, *Physiology and Biochemistry of Extremophiles* (n. 8).

27. A. E. Walsby, *Trends in Microbiology* 13, 193–5 (2005).

28. Here's the calculation. Begin with a square cell with dimensions 10 × 10 × 1 unit. This has a surface area of 240 units^2, volume of 100 units^3, and surface area to volume ratio of 2.4. The same cytoplasmic volume can be packaged into a sphere with a radius of 2.9 units, surface area of 104 units^2, and surface area to volume ratio of only 1.04 (= 3/radius).

29. C. Gostinčar et al., *FEMS Microbiology Ecology* 71, 2–11 (2010).
30. <http://www.oecd-nea.org/rp/chernobyl/co5.html>; the Gray is the SI unit for absorbed radiation dose. One Gray is defined as one joule of energy, in the form of ionizing radiation, absorbed by one kilogram of matter.
31. N. N. Zhdanova et al., *Mycological Research* 104, 1421–6 (2000).
32. N. N. Zhdanova et al., *Mycological Research* 98, 789–95 (1994).
33. N. N. Zhdanova et al., *Mycological Research* 108, 1089–96 (2004).
34. E. Dadachova et al., *PLoS ONE* 2(5), e457 (2007).
35. P. Huyghe, *The Sciences* 16–19 (July/August 1998).
36. J-I. Kim and M. M. Cox, *PNAS* 99, 7917–21 (2002).
37. S. B. Pointing et al, *PNAS* 106, 19964–9 (2009). Organisms lacking UV protection can survive on the underside of rocks and are described as *hypolithic*.
38. High concentrations of toxic perchlorate on the Martian surface are one of the many challenges for a manned mission to the red planet. Perchlorate is a component of solid rocket fuel, fireworks, and explosives. Research on terrestrial bacteria and archaea that break down perchlorate has led to the suggestion that this compound could support microbes beneath the frozen Martian surface. *Archaeoglobus fulgidus* is a hyperthermophilic species of archaea that reduces perchlorate: M. G. Liebensteiner et al., *Science* 340, 85–7 (2013).
39. C. D. Parkinson et al., *Origins of Life and Evolution of Biospheres* 38, 355–69 (2008).
40. C. Humphries, *Science* 335, 648–50 (2012).
41. P. Zalar et al., *Fungal Biology* 115, 997–1007 (2011).
42. J. A. Littlechild, *Biochemical Society Transactions* 39, 155–8 (2011). Z. E. Wilson and M. A. Brimble review carbohydrates, lipids, and secondary metabolites from extremophiles with a variety of biological activities in *Natural Product Reports* 26, 44–71 (2009).
43. K. B. Mullis, *Scientific American* 262, 56–61, 64–5 (1990).
44. M. R. Tansley and T. D. Brock, *PNAS* 69, 2426–8 (1972).
45. M. M. Littler et al., *Science* 227, 57–9 (1985).
46. L. A. Levin, *Palaios* 9, 32–41 (1994).
47. R. J. Richards, *The Tragic Sense of Life: Ernst Haeckel and the Struggle over Evolutionary Thought* (Chicago: The University of Chicago Press, 2008).
48. J. Pawlowski et al., *Journal of Eukaryotic Microbiology* 50, 483–7 (2003).
49. *Valonia* is a member of the green algal order called Siphoncladales; *Caulerpa* and *Halimeda* belong to the Order Bryopsidales.
50. *Epulopiscium fishelsoni* is another large bacterium. It lives in the gut of a surgeonfish and its dimensions are 200–700 μm x 80 μm; E. R. Angert, K. D. Clements, and N. R. Pace, *Nature* 362, 239–41 (1993).

51. L. Selbmann et al., *Studies in Mycology* 51, 1–32 (2005).

52. L. G. Sancho et al., *Astrobiology* 7, 443–54 (2007); J. Raggio et al., *Astrobiology* 11, 281–92 (2011).

53. J-P. de Vera, P. Rettberg, and S. Ott, *Origins of Life and Evolution of Biospheres* 38, 457–68 (2008).

54. G. Horneck et al., *Advances in Space Research* 14, 41–5 (1994). NASA's Long Duration Exposure Facility was launched on a Space Shuttle in 1984 and recovered in 1990. It was scheduled for retrieval after 11 months in space, but the mission was extended following the Space Shuttle *Challenger* accident.

CHAPTER 8

1. S. Naeem, J. E. Duffy, and E. Zavaleta, *Science* 336, 1401–6 (2012).

2. M. G. A. van der Heijden, R. D. Bardgett, and N. C. van Straalen, *Ecology Letters* 11, 296–301 (2008).

3. M. G. A. van der Heijden et al., *Nature* 396, 69–72 (1998).

4. S. A. Schnitzer et al., *Ecology* 92, 296–303 (2011).

5. G. W. Griffith, *Trends in Ecology and Evolution* 27, 1–2 (2012).

6. T. Curtis, *Nature Reviews Microbiology* 4, 488 (2006).

7. D. L. Stokes, *Human Ecology* 35, 361–9 (2007).

8. M. J. Costello, R. M. May, and N. E. Stork, *Science* 339, 413–16 (2013).

9. Tom Curtis of blue whales and pandas fame, *Nature Reviews Microbiology* (n.6), used this reasoning to consider the scale of the challenges in microbial ecology. Analysis of sequences obtained from seawater samples from the English Channel and a global archive of sequences from multiple marine sites showed a lot of overlap; S. M. Gibbons et al., *PNAS* 110, 4651–5 (2013). This finding supports the idea that similar microorganisms exist in all locations and bloom according to environmental conditions. It follows that deep sequencing of seawater collected from any location could reveal most of the microbial diversity throughout the earth's oceans. Statistical analysis of the two data sets suggested that a comprehensive inventory of marine prokaryotes might be obtained by amplifying 200 billion sequences from more than 200 liters of seawater. Deep sequencing of this kind might be feasible in a few years.

10. S. A. Amin, M. S. Parker, and E. V. Armbrust, *Microbiology and Molecular Biology Reviews* 76, 667–84 (2012); R. Stocker and J. R. Seymour, *Microbiology and Molecular Biology Reviews* 76, 792–812 (2012).

11. T. Pradeu, *The Limits of Self: Immunology and Biological Identity* (New York: Oxford University Press, 2012).

Color Plate Credits

1 Shutterstock.com/ Lebendkulturen.de
2 J. Leidy, U.S. Geological Survey of the Territories Report 12, 1–324 (1879).
3 Phillipe Crassous/ Science Photo Library
4 www.istockphoto.com/© NNehring
5 Phillipe Crassous/ Science Photo Library
6 Eye of Science/ Science Photo Library
7 Reprinted from Deep Sea Research Part II: Topical Studies in Oceanography, Volume 58, Issues 23–24, 1 December 2011, A.J. Goodaya, A. Aranda da Silvab, J. Pawlowski, Xenophyophores (Rhizaria, Foraminifera) from the Nazaré Canyon (Portuguese margin, NE Atlantic), page 2407, fig 7, Copyright 2013, with permission from Elsevier
8 Shutterstock.com/ Lebendkulturen.de
9 Dr. Peter Siver, Visuals Unlimited / Science Photo Library
10 Photo: Professor Timothy James
11 Eckhard Voelcker
12 © CCALA Culture Collection of Autotrophic Organisms, http://ccala. butbn.cas.cz
13 Russell Kightley/ Science Photo Library
14 From Untangling Genomes from Metagenomes: Revealing an Uncultured Class of Marine Euryarchaeota by Vaughn Iverson et al. Science 335, 587 (2012); DOI: 10.1126/science.1212665. Reprinted with permission from AAAS

INDEX